Tone

The Electric Revolution

THE ELECTRIC REVOLUTION

by R. A. S. Hennessey

ORIEL PRESS

First published 1972

ISBN 0 85362 133 0

Library of Congress Catalogue Card No. 77–166002

Published by
Oriel Press Ltd
32 Ridley Place, Newcastle upon Tyne
England, NE1 8LH

Text set in 12 on 14 pt Baskerville

Printed in Great Britain by
Clarke, Doble & Brendon Ltd, Plymouth

Introduction

THE secondary fuel industries—gas and electricity—have largely escaped the attention of historians, in a way that transport, for example, has not. If this is because they seem in any way less interesting or important, or even less 'romantic' than railways, steelmaking, coal mining or grand engineering, then the poverty of histories on the subject of electricity supply is based on a profound misconception; a matter which this book attempts to rectify.

This work describes the 'Electric Revolution', 1880–1930. In those fifty years a new industry grew from the most obscure and makeshift beginnings into a position whereby it was a major economic fact, and an important foundation for society and civilization. The broad outlines of this saga, adorned with such details as will furnish proof for general assertions, are as entertaining and important as any economic history. Because electrical engineering has already been admirably described elsewhere, there is a slight bias here towards describing electricity generation, and its history, for which the author considers no apology necessary. This vital industry (and the adjective is used advisedly) has almost escaped record, and R. H. Parsons' classic work[1] on it formative years is now hard to obtain.

The term 'Electric Revolution' is a convenience, not an attempt to create yet another economic concept. But what other word could describe the speed and excitement manifest by the electrical industry in the onrush of its early development? Its practitioners were imbued with intense enthusiasm, and they presided over sweeping changes in the energy-base of civilization; they struggled with opulent vested interests and—like Ferranti, Crompton or

Merz—saw victory and achievement in their own lifetimes. Some may demur at calling it a revolution, but it was nothing short of one.

NOTES

1 R. H. Parsons : *Early Days of the Power Station Industry*, C.U.P. 1939.

Contents

Acknowledgments

I should like to thank the following for the help that they have given in the preparation of this book: the *Architectural Review*; the British Electrical and Allied Manufacturers' Association; British Railways; the Central Electricity Generating Board; Chester Public Library; the Electrical Association for Women; the Electricity Council; the *Electrical Review*; the Electricity Supply Board of the Republic of Ireland; the Federal Power Commission, Washington DC; Messrs Ferranti; the General Electric Company; T. G. N. Haldane, Esq; G. S. Hearse, Esq; Captain S. J. Hennessey, OBE, RN; Messrs Hoover; The Institution of Electrical Engineers; the Literary and Philosophical Society of Newcastle upon Tyne; Messrs Merz and McLellan; Lord Newborough; the North Eastern Electricity Board; the North of Scotland Hydro-Electric Board; Rochester, NY, Public Library; the South of Scotland Electricity Board; South Shields Public Library; the States of Guernsey Electricity Board; the Trades Union Congress.

Illustrations: *Architectural Review*, 33, 38; British Rail, 23; CEGB, 14; Chester Public Library, 18; *Electrical Magazine*, 7, 29; *Electrical Review*, 1, 2, 4, 5, 6, 8, 10, 11, 12, 13 (Kettering), 19, 24, 31, 32, 34, 35, 36; Electricity Supply Board, Republic of Ireland, 17; Ferranti, 21; GEC-AEI-EE, 39; GEC of America, 26; *Illustrated London News*, 15; Institution of Electrical Engineers, 22; London Transport, 30; Messrs John Murray for quotation from Sir John Betjeman's *The Metropolitan Railway*, 1; South of Scotland Electricity Board, 16; South Shields Library, Parry Collection, 3, 13 (Weardale), 27; Sir Kenneth Swan, 20.

A Note on Abbreviations

1. Current is marked as AC (alternating current) and DC (direct current). Before the Great War it was common to call DC 'continuous current' (CC) but this work holds to DC in all cases.

Frequency of AC is marked as Hz (Hertz) which used to be called cycles per second, and the usual abbreviations regarding output are employed: kW (kilowatts = 1,000 watts); kWh (kilowatt-hours); MW (megawatts = 1 million watts). Voltage, likewise, is usually marked as kV (kilovolts).

2. Prices are marked in current terms. To adjust to constant terms one may assume (approximately) that £1 of 1900 = £8 of 1971. British currency was decimalized in 1971 so the decimal equivalent is indicated, where confusion might arise, when pre-1971 prices are quoted. Old pence (pre-1971) are given thus—5d and new (decimal) pence are given—5p.

3. Electricity suppliers were of various distinct types. An 'Undertaking' usually referred to an institution (private or municipal) which supplied electricity to consumers, and the term implied that it had parliamentary powers to do so. If it did not, it would be an 'unauthorized undertaker', or a 'non-statutory' one; the Electric Lighting Act of 1909 sections 23 and 25 cover this. A 'power company' supplied 'bulk' electricity to undertakers, but did not necessarily sell it to the public unless authorized to do so by statute (see Chapter 4). Undertakers might generate and distribute power, some merely bought it (i.e. from a power company, or a generating undertaking nearby) and retailed it, without owning a power station of their own.

Part One

1, *The Electric Revolution*

THERE has been a surfeit of 'revolution' categories in economic history; D. C. Coleman reminded and warned us[1] to use the term 'Industrial Revolution' with rigour and to confine its use to that period when society and the economy were revolutionized, in the late eighteenth and early nineteenth centuries. After that period, the railway and automation 'revolutions' have merely accelerated the pace and intensified the effects. They do not constitute discrete revolutions, nor does the term 'Electric Revolution' attempt to suggest the existence of a new and hitherto undiscovered period. It refers to certain events *circa* 1880–1930. In those fifty years the electricity supply industry grew from a few portable arc-lighting systems into a network of some five hundred power stations; a few of them (like Battersea or Barking) constructed on the grand scale, and capable of powering large tracts of urban civilization. Electrical engineering had become a major industry, and the engineers were organized into professional and trade associations, formalized and accepted as normal. These things were remarkable developments in themselves, little short of revolutionary.

Yet there must be added to this internal development the effects of electricity on economic and social life. By 1930 there were nearly three million consumers of electric power. Much public and private lighting had become electric; electric trams, cinemas, dentists' drills, cranes and radio sets were evidence of the widespread effects of electricity in many aspects of life. If power, with food, is a pillar of civilization, then the ability to tap larger quantities of it is bound to alter the foundation and—to borrow a Marxist term—the 'superstructure' of society. This fact, added to the remarkable growth of the industry itself, is again nothing less than revolutionary. Like the railways (only more so) electricity was born in one world, and matured in another, the latter being very much a product of the industry in question.

Thus 'Electric Revolution' is a convenient term which encompasses a mass of integrated changes which occurred between 1880 and 1930.

As with the seasons of the year, so too industries live and die in cycles. Some enjoy a long innings, seemingly an eternal one, like viticulture or papermaking. Others enter an unmistakable secular decline, perhaps at a great age (like coal mining), or a shorter one, like the cinema. Their strength or influence in the past cannot avert the inexorable 'destiny of cycles' as Spengler might have it. Other industries have perished completely, giving employment only to economic historians, and joy to archaeologists.

When an industry is young, or in its 'springtime', it affects certain marked and recognizable characteristics. It is ebullient, lively, aggressive, and it experiments readily. Bright and unorthodox young men rise rapidly in its service; it manifests a refreshing, brash and slightly naïve self-confidence, above all it explores many, perhaps most, of its possibilities. There are examples of this in recently recorded history: railways and electricity. The confident and rollicking enthusiasm of *Herapath's Railway Journal*, the early railway chronicle, is matched by that of the infant *Telegraphic Journal* (later *Electrical Review*) or *Electrician*. Bold spirits, like I. K. Brunel and Joseph Locke have their parallels in men like Charles Merz or Sir Charles Parsons. The battle of the gauges[2] on the railways was like the battle of the standards, or systems (alternating current *versus* direct current) in the power industry: both were functions of furiously confident faiths, amounting almost to a theological controversy or a religious war.

Later, the industry becomes established and professionalized. It supports a galaxy of professional institutions and academies; but it also becomes orderly and rather duller. The gauge war on the railways was followed by the Gauge Act, 1846; the warring medley of electric systems was steadily rationalized after the Electricity Supply Act, 1926. By 1850 trains were commonplace; by 1920 few people in Great Britain boggled in wonder at an electric light. The magic of novelty dies away, and a certain stodginess replaces it. Fine traditions of public service and reliability may well be established, but the stimulating amateurism has gone for ever. The hackneyed lines of Wordsworth have relevance when one senses, through old periodicals or memoirs, the hot optimism and boyish confidence that abounded in the early Electric Revolution; hard to recapture in our rather jaded and blasé times:

Bliss was it in that dawn to be alive
But to be young was very heaven.

The dawn of the electric age—the palaeo-electric era—was about a half-century. Without being dogmatic or exclusive one would suggest 1880–1930 as its confines. These limits could be symbolized by the rather ineffectual Electric Lighting Act (1882) which at least showed that the new energy was worthy of statutory recognition and control, and at the other end, the erection of the last grid tower[3] in 1933, marking the completion of the original (132kV) national grid. By that date, electricity had come of age in Great Britain.

Not all vestiges of the dawn had been dispersed by 1930, of course. The enormous and confused system of electricity generation and distribution lingered on, or rather strode on (for it was no ailing colossus), until nationalization in 1948. Until then it was the largest and most capital-intensive remnant of the golden age of municipal enterprise, civic pride and particularism. After 1948 (save for Scotland) London increasingly proposed and disposed in matters electrical. Once again the provinces were to be subordinated to Megalopolis, whereas in the Electric Revolution, power generation was unashamedly, aggressively and often very inefficiently, local.

By 1970 most of the vintage power stations have gone, in the same way that very few copper-bowl electric fires, *art nouveau* electroliers (see figure 1) and Egyptianesque 'nickelodeon' cinema façades are left. In its comparatively short life, the electricity industry has sustained other industries-entire that have suffered remarkably brief life spans of their own: electric tramways and trolleybus systems are virtually all gone, and the cinema industry has shrunk sharply since 1945. Meanwhile, the actual electric power supply industry itself has become massive and vigorous. To describe the 10,000 miles of pylon-borne circuits as our industrial and social 'arteries' is now a cliché, and like most such clichés, unarguably true.

In the early electricity age, the electrical engineers and salesmen were on the march against the entrenched hordes of coal, steam and gas. They found their way made difficult by statutory obstacles, fashioned by crafty politicians and the 'unscientific' interests that they served. Still, they moved on, brimful of confidence and always looking for new fields upon which to do battle with orthodoxy and reaction. Contemporaries of many types said, or did, things which reflected their faith in the new Prometheus: to Lenin, Communism was 'Socialism plus electricity'. In 1919, in the USA, *Technocracy Inc* was founded by Howard Scott *et al.*, to 'reshape the North American Republic around the dynamo'; the perceptive Henry Adams (1900) knew that 'the age of the dynamo' was upon us, 'the period . . . is in full swing, and, gee-whacky! how it is going!'[4] This dynamo-based civilization (or to be pedantically accurate, alternator-based one by 1920) has not brought quite what was expected by the more uncritical electrophiles of the Edwardian age. Significantly, Eugene O'Neil (1929) called the work in which he delved into the 'failure of science and materialism to give any satisfying new (God)' *Dynamo*. The terrifying dependence of a civilization based on generated power renders works like E. M. Forster's *The Machine Stops* highly discomforting. Henry Adams did end his speculation on the Age of the Dynamo: 'I . . . go down to the Champ de Mars and sit by the hour over the great dynamos, watching them run as noiselessly and smoothly as the planets, and asking them—with infinite courtesy—where in Hell they are going.'

If Adam's observations could be dismissed by the young technocrats, as the views of a soured Boston brahmin, they did display, nevertheless, that by no means all important opinion was carried along by the propagandists of electricity and science-based civilization. However, to many ordinary people, and certainly to the growing ranks of young electricians, the future of the industry was bright, and so should be the rational society based on it. Typically, not a few of these electricians were Fabian-type socialists. Many people had strong faith in the new power, and might forecast of AD 2000 (like E. F. T. Marinetti in *La Futurisme*, 1912) that we would be 'extremely revivified, shaken and tamed by new electrical energies'; like the Marquess of Shropshire in John Galsworthy's *A Modern Comedy* (1929) which continues the *Forsyte Saga*. Lord Shropshire 'was dictating to his secretary a letter to his County Council, urging on them an item in his lifelong programme for the electrification of everything. One of the very first to take up electricity, he had remained faithful to it all his brisk and optimistic days.' There was, such men claimed, nothing that electricity might not do, and it was all for the best.

Within fifty years, the Electric Revolution had run its course. Electricity had passed from being the power of the future to being the power of the present—with an ever-expanding future unfolding before it. As well as occurring in an age of rapid social and technological change, the Electric Revolution substantially accelerated both. It straddles, as no other body of technology has done, a gulf between distinct societies. Given the increasing standardization of the 'global village' in which we now live, it is hardly likely that any future change will bridge so wide a difference in so short a time. The culture in which Swan and Edison, the inventors of the electric light, grew up is so remote to us as to seem part of another planet, but the England of the 1930s has a distinctly familiar feel about it, for it was strongly based on electric power. Its entertainments, communications, industrial structure were already heavily dependent on electric power in a way that no former age had been. There were other signs, too. The centralized and totalitarian states that developed before the Second War owed not a little to electrically based methods of communication and living. Many of these techniques remain, and are quite likely to furnish a basis for absolutism in the future. Data-banks, 'bugging', propaganda, the great metropolis of high-rise blocks and linear induction transport will all depend upon that power welcomed in an open and sanguine way by the optimists of 1880–1910. They accepted all change as *ipso facto* for the better, by their standards. When other forces seized upon and used electricity they were less sure. By extending us, electric power has extended the scope of the good and evil that works through us; it could never have been otherwise.

NOTES

1 D. C. Coleman, *Industrial Growth and Industrial Revolutions*; Essays in Economic History Vol 3, ed E. M. Carus-Wilson.

2 The standard gauge of 4 ft $8\frac{1}{2}$ in, *versus* the broad gauge of 7 ft $0\frac{1}{4}$ in favoured by I. K. Brunel and the Great Western Railway.

3 Near Fordingbridge, Hants—see figure 10.

4 H. Adams, *The Education of Henry Adams*.

2, *Fiat Lux*

In the centre of Newcastle upon Tyne there is a massive, grey administration building—Carliol House (1927). It contains the headquarters of the North Eastern Electricity Board, and when it was opened it housed the centre of the great NESCO electric power web,[1] and also the offices of Merz and McLellan, the consulting engineers. It is fitting that this building should display a bronze plaque in honour of Sir Joseph Wilson Swan, inventor of the electric light bulb, who was born at Sunderland in 1828. The tablet's message ends simply: *Fiat Lux*; let there be light.

In its first years as an industry, electricity supply was seen as a lighting agent almost exclusively. Later it was used for traction, then in industry and finally to perform various useful domestic chores. But above all, light—a fact enshrined in the names of many electricity companies before nationalization in 1948, like the Peacehaven Electric Light and Power Co, etc. In the USA, the light's co-inventor, Thomas Alva Edison is similarly honoured in the names of present-day power giants, like Con Edison. The lamp was the foundation of the electricity industry in another way. The profits from early lamp patents, and later from electric lamp price-rings, were the single greatest support for electrical engineering companies in many countries. This was very much the case in Great Britain, where the 'infant Hercules' of electricity was born in a very hostile world of vested power interests, industrial inertia and unsympathetic political attitudes. The lamp was the sure defence behind which the industry grew.

The need to improve the means of generating power for lighting in turn stimulated important developments in electric engineering, as traction was to do later. By 1900 many—perhaps most—of the principal breakthroughs in 'basic electrics' had been made; the position was thenceforward consolidated and perfected. This is not to discount such great advances as semi-conductors, non-mechanically generated power, linear induction, etc. Yet the burst of creative energy towards the end of the nineteenth century, from men like Kelvin and Tesla, is staggering by any standards. It was a flowering of talent that launched the Electric Revolution and gave it abiding momentum. By 1900 light, intellectual as well as incandescent, blazed most fiercely.

The primeval electric light was known in the very early nineteenth century. Davy had produced an 'arc light' by passing an electric current into two charcoal rods, whose tips were barely separated. The carbon arc light which resulted was brilliant, and suitable for lighting public places. As such it overlapped the true era of the electric bulb or lamp with which we are familiar. The Victorians called this bulb the 'incandescent' electric light in order to distinguish it from the arc. The arc light was eventually displaced by mercury- and sodium-vapour lights, or very powerful bulbs when it was wished to light highways, sports arenas etc. *en grand*. The arc's main drawback was the short life of the carbons, which needed frequent replenishing. Such a disadvantage was not insuperable in an age of cheap labour, especially where arc lights were clearly superior for the job in hand. They were used, for example in the South Foreland lighthouse (1858), and Souter Point light (1867). In the late 1870s the improved dynamo led to the development of simple dynamo-and-arc sets, such as the popular model produced by the American pioneer Charles F. Brush. A variant of the standard arc was the 'Jablockhoff candle' (1876), which placed the carbons parallel, but embedded them in a coat of china clay. They were compact and easy to handle, but had severe operational disadvantages. They could not, for example, be re-lit if a gusty wind or power cut broke the arc.

The Electrician reporting on the various arc systems (1881) then being adopted by the primitive lighting industry observed, rather heavy-handedly: 'The Jablockhoff light seems to remain at rest and is, we suppose, thankful.' Only three years before, the same periodical was complaining that while the electric light of all types was popular in Paris, 'yet in London, it was nowhere to be seen'.

By the end of 1878, however, the arc light had appeared in London, like 'half a dozen harvest moons shining at once in the Strand'—illuminating the Gaiety Theatre. There were a few other late-1878 systems as well as the Gaiety's enterprise: Victoria Embankment, Billingsgate Fish Market and Holborn Viaduct. The Embankment system used Jablockhoff candles, and extended about six thousand feet each side of old Waterloo bridge (on which a further ten lamps were placed), but the agreement between the Metropolitan Board of Works (ancestor removed of the Greater London Council) and the Jablockhoff Electric Lighting and Power Co lapsed in 1884 and was not renewed. Gaslight was restored, as it was to the Holborn Viaduct experiment (late 1878–May, 1879) because arc lights, although very bright, were four times as costly as gas lights.

When good light was essential, not merely prestigious, then arc lights were sometimes adopted more permanently. By 1880 we find them in Woolwich Arsenal, W. D. & H. O. Wills' tobacco factory at Bristol and the Trafalgar Colliery in the Forest of Dean. In the sphere of entertainment, it is interesting to note the arc light illumination of the promenades at Westgate-on-Sea and Blackpool, and the first illuminated football match:

> The intense interest aroused by the application of the electric light to novel uses was strikingly apparent on Monday night in Sheffield, when nearly 30,000 people gathered at Bramall-lane grounds to witness a football match played under that light. The match, which was played by two teams belonging to the Sheffield Football Association commenced at half past seven o'clock. The electric light was thrown from four lamps,

thirty feet from the ground, and the rays, which were of great brilliancy, lighted nearly the whole of the ground, the players being seen almost as clearly as noonday. The brilliancy of the light, however, dazzled the players, and sometimes caused strange blunders... (*The Electrician*, 1878).

These early arc systems were all small and compact, usually run off dynamos which might be mounted on, or driven by, traction engines; nearly all the examples alluded to were of that type. The arc light impressed contemporaries deeply, and induced (rather early in the day) complacent assumptions as to its future. Hints of the possibility or desirability an altogether smaller and more convenient design for domestic use, and moreover one that did not eat up carbons, were ignored or pooh-poohed by the pundits. Thus Sylvanus Thompson, a leading academic and engineer: 'but this I can tell you as the result of experience, that any system of lighting by incandescence will utterly fail from the economic point of view . . .'[2] Other leaders of technical opinion, including Hopkinson and Siemens did not even rate the incandescent lamp a mention when reporting to the Playfair parliamentary committee which was investigating the question of electric light in 1879.

The tendentious debate as to who invented the electric light is largely over now. Honours are usually accorded jointly to Swan and Edison. Although Swan was strictly the first, so great and so close was Edison's achievement that it would be pedantic to exclude him. On the other hand, Swan worked full time as a chemist, doing his work on the light in spare time; Edison had the ample help of his full research establishment at Menlo Park, New Jersey. Swan held that he had developed other men's pioneer work, notably that of Staite, who had electrically excited iridio-platinum wire during a lecture at the Athenaeum, Sunderland. This was observed by the youthful Swan (1845). As so often, the 'first' inventor is a rather arbitrary text-book decision which serves to fix our attention on the date when an innovation became technically and commercially feasible.

Edison and Swan both suffered many setbacks on the way to finding a workable and commercially attractive lamp. The single greatest help for their efforts was the Sprengel pump, which could evacuate air from glass bulbs, allowing the electrically excited filament within to glow, not burn. On 18 December 1878 Swan demonstrated such a bulb to the Newcastle Chemical Society, although it burnt out after a few minutes because of excessive current. However, the point had been made in public. Swan was soon a *cause célèbre*, and toured the kingdom making lectures and demonstrations. On a famous occasion, when he was lecturing to the Literary and Philosophical Society of Newcastle, and addressing a large galaxy of the Northern intellectual, commercial and social élite, he made a deep impression by having seventy gas jets turned down, and their job taken up instantaneously by twenty electric bulbs (20 October 1880).

The incandescent light had been proven. Now it remained to install it for routine domestic use, and to demonstrate its superiority for such a purpose. An arc light was too brilliant for the home, the electric light had to be 'subdivided' as Swan's contemporaries put it. Swan lit his own home, and that of his friend Sir William Armstrong with complete success. The more go-ahead entrepreneurs of the early 1880s seized upon this chance to get better light and good publicity into the bargain. However, in the early days of electric lighting it was only the very rich, or the managers of large establishments who could afford it, hence Sylvanus Thompson's premature remark. Bulbs cost twenty-five shillings each (£1·25) and current was about one shilling per unit (5p). At the same time arc lamps cost £15, and both systems in those early days needed their own generating sets as well. In a letter to *The Times*, a Mr Octavius Coope calculated that 'gas plant was cheaper to install in his home (£1,333 to £1,470 for electricity) but that electric running costs were half those of gas, and he had 'no nuisance of lime or of tar and other refuse; no leakage of gas, no smell . . . or damage to my garden'.

A quickly growing market encouraged mass production and the division of labour in the classic way of Adam Smith. Swan founded the Swan Electric Lamp Co with a small factory at Benwell, near Newcastle. German glass-blowers had to be encouraged to go there from Thuringia, for no British glass-blowers had the necessary skill. After reorganization, and the settling of outstanding differences, Swan joined forces with Edison and they launched the 'Edison and Swan United Electric Light Co' (1882); their lamps were later and more conveniently known by the portmanteau trade name 'Ediswan'.[3] By 1883 Ediswan was making ten thousand lamps per week, and was able to reduce unit costs and prices. About this time Swan moved down South, to Bromley in Kent. Here he was near to London, the fastest growing sector of his market. A market-oriented factory was built at Ponders End, and the epic of the electric lamp was thereafter ever onwards and upwards.

Other manufacturers later established themselves (especially after the patents relaxed in 1893) and took out patents for their own improvements on the Ediswan archetype lamp. The main advances were Siemens' tantalum filament bulb, and the tungsten filament gas-filled bulb, the GEC 'Osram'.[4] Most of the large makers' names or brand names were household words by 1930 (see figure 4). In addition, gas-filled 'fluorescent' tubes for advertising came to Great Britain from America *c* 1900. They were called 'Moore Tubes', and had a short and unreliable life. The superior neon tube came in 1913, first being used to adorn the façade of the West End cinema in Coventry Street, London—in the tradition of the Gaiety back in 1878. The adoption of sodium lights (1932) and mercury vapour lights (1947) replaced the last of the arc lights. All these changes freed people more than gas or oil had ever done from the 'thraldom of darkness'. Within Great Britain another great change is exemplified by Swan and his achievement, that is his move to Bromley and Ponders End. This is one of the first examples of the 'drift to the South' of a market-oriented industry. This drift was to accelerate and damage the North badly, especially after the Great War, and it was a movement largely made possible by the ample supplies of electric power down South.

Economic and social changes have occurred so rapidly in the electric age that there are still (1971) people who recall the thrill of their first experience of electric light installations. An old lady wrote thus, recently:

> In 1889 we stayed at an hotel in Southampton, that had electricity installed, and our mother (who wasn't sure how to switch on the light) had to ring for the maid, to ask her how to do it. In 1890 we were living in a small town in Yorkshire (Harrogate), and ours was the first home in the town to be lighted with electricity. We generated our own power. People living in the town used to drive up to our gates in carriages and also bring their visitors, to see 'the house with the lights'. As young children when we heard the sound of horses' hooves at dusk in our quiet road, we rushed around the house switching the lights on and off so that the spectators outside would not be disappointed.[5]

The first shop to use the new light for publicity was Swan's own (Mawson and Swan, Moseley Street, Newcastle) and he was soon copied by a nearby draper, James Coxon. In June 1881 the Inman liner ss *City of Richmond* was the first ship to mount electric lights. An even more spectacular display was D'Oyly Carte's illumination by electricity of the Gilbert and Sullivan opera *Patience* by 824 Swan lamps, with a further 370 in the auditorium (1882). The *Electrical Review* sent a reporter who noted, a little acidly, 'we noticed on the stage a considerable number extinguished—in some rows as much as ten per cent of them merely glowed like *red* hot pieces of charcoal'. D'Oyly Carte thought highly enough of them, however, to similarly light *Iolanthe* later, and to open up new electric possibilities by tipping the fairy queen's wand with a glow lamp that could be switched on or off; the rank-and-file fairies had mini-lamps in their hair, and they 'glittered like a swarm of fireflies', no doubt delighting the Victorian audience, although wildly unsafe by later standards.

By 1885 a 'small incandescent lamp' was used to explore the larynx of the US ex-President General Ulysses S. Grant, who was suffering from a throat ailment. The London and North Western Railway's steamship *Kathleen Mavourneen* had been 'fitted up in the most approved manner, and lighted throughout by the electric light', while the Wellington Club, Liverpool, ballroom used just over one hundred Swan lamps in lieu of five hundred and fifty wax candles, the Swan lamps mounted 'inside elaborate glass chandeliers' producing 'very striking and beautiful effects'. Swan's erstwhile helper and subordinate, R. E. B. Crompton (see p 108) struck out on his own and contracted to light the Mansion House, London, and St Enoch Station, Glasgow. The youthful S. Z. de Ferranti (see p 109) set about lighting Cannon Street station in London. Many of the great electrical engineers and contractors of later years had their 'baptism by light', for light was then the sole purpose of electricity.

Edison's hold on electric light patents was the sure foundation upon which the giant General Electric Co of America (no relation to the British GEC) was built in 1892. It was the result of a merger between the Edison General Electric Co, and the Thomson-Houston Co. In this way also, electric lighting was to be a foundation of the entire electricity industry. In Great Britain the lamp-makers formed into price rings early in the twentieth century (the British Carbon Lamp Association). The pioneer, Ediswan, rather fell behind its rivals technically, British Thomson-Houston, GEC and Siemens, all of whom based their success on tungsten filament lamps (Tungsten Lamp Association, of 1912). In 1919 the two rings joined and became the Electric Lamp Manufacturers Association (ELMA), and five years later ELMA was party to the international lamp agreement, 'Phoenix'. This cartel was an economic classic: it completely dominated the market for a demand-inelastic product; it could squeeze handsome profits out of the public. A government report[6] of 1920 showed that standard three shilling (15p) lamps could easily sell for two shillings (10p) and still give a comfortable profit. Yet the British electrical engineering industry was often weak between the wars, and the lamp ring supplied the profits which took care of the overheads. It was broken up in the 1950s by government order. Throughout the Electric Revolution the steady stream of revenue from lamp sales supplied the lifeblood for British electrical engineering, so that Swan's genius founded the electricity industry in more ways than the obvious one.

The Electric Revolution urged men to explore the possibilities of the new power, often in the hope of emulating Swan or Edison, and striking a rich vein. Many of their discoveries have become everyday devices, others are merely quaint illuminants of dusty archive searching. One long lived product of the early days is electric heating. Early heaters (see figure 2) mounted special bulbs (sometimes red or orange tinted) in front of reflectors. In January 1895, the Vaudeville Theatre in London was worried about the effects of a cold snap on their audience ratings; the manager asked Crompton for help. He received this call at 11 am, and by 6 pm had put in a series of radiators which kept the cold at bay. The year before, Crompton had fitted up Miss Fairclough's School of Electric Cookery (Gloucester Road, London) which establishment offered instruction in the newest domestic art. Crompton, and the City of London Electric Lighting Co were interested in pushing the new cookery because it would help to spread the load on power stations. Crompton, in fact, invented the term 'load factor' (1891) to describe the relationship between the average, and the maximum possible, load on generating machinery. Since it was often as low as 10 per cent in those days, anything like electric cooking was a very welcome burden in off-peak hours. The City of London ELCo ovens cost £7 10s 0d to £14 0s 0d (£7·50 to £14) or they could be hired at seven shillings to twelve shillings (35p–60p) per quarter—an early example of this familiar way in which power industries encourage the purchase of load-widening outlets. Another *douceur* to encourage custom was the installation of a special oven circuit for power at 4d per unit—half the lighting price. The City of London Electric Lighting Co was an AC distributor, hence its acute interest in easing out the daily load. Alternating current is unable to be stored, so its generating machinery must be capable of meeting any demand. Under such circumstances low load factors are very wasteful. Crompton, a DC man, backed a system where storage was possible in batteries, but he backed DC (among other reasons) because generators could always be working at near maximum output, either charging batteries for the next 'peak' or helping to ease it when it came; domestic ovens helped improve the load factor even more. By 1900 there were, for example, electric cigar lighters, curling tongs, irons and kettles for sale. One likes to think of an Edwardian merchant-prince, velvet clad and ensconced in his panelled library, unwittingly doing his bit towards easing the load factor problem by electrically lighting his corona-corona, late at night.

There were some other and big users by 1900 also, above all tramways (which were often backed by electricity suppliers for load factor reasons, see p 15) and the early cinemas. By 1914 more gadgets were on the market, notably the vacuum cleaner;[7] there were also a few fringe lost causes, always a part of any revolution. 'Electroculture' of plants by pushing naked terminals into the soil near their roots enjoyed little success, although electricity helped agriculture in many other ways later. A semi-starter was the electric tractor, running off a long lead, which was rather dangerous and did little to bridge the gap between the massive steam traction engine and the horse, a gap filled in due course by the motor tractor. The *Electrical Review* fought a successful court case after lambasting a quack who sold 'electric belts' for rheumatism. The periodical was out to keep the name of electricity clean, but the doctor was one of many electro-cranks who played with fire. Horrors indeed at a report in the Newcastle *Evening Chronicle* (13 November 1912) which described the death of 'Mr Moreton Thompson, a solicitor, in an electric bath' which drew attention 'to the question whether the appliances in some baths are installed in such a way as to secure safety'. An expert opined: 'with resistances shunted from the mains, electric baths [are] dangerous, and accidents are liable to occur . . . the chances are that Mr Thompson did not get a little of the current but all of it. A man in a bath is a splendid conductor. He puts his hand on a tap and that is enough . . . the only way to avoid such accidents is to have a separate dynamo. This would make electric baths safe.'[8] The Electric Revolution taught people the facts about the new power the hard way, and the rigorous safety standards of later years are largely the result of the revolutionary martyrdoms of such men as Mr Thompson.

Militarists saw possibilities in electric power very early, largely as a lighting agent. Searchlights were used to good effect in the South African War (1899–1902) and the Russo-Japanese War (1904–5). Electricity had long helped telecommunications in peace and war (largely battery powered, however)[9] and in the period when any and every road was being explored there was much talk of electric weapons. One was a suggestion that electricity could be projected at immensely high voltages through the air, to drive machinery, warships or missiles.[10] The great Tesla worked on this, and many years later it was hinted that he had outlined a type of death ray. Perhaps he succeeded, but feared that civilization could not contain such fire of the gods.

By 1914, electricity had conquered many old areas of enterprise, and had opened up new ones of its own. What started as a lighting power was driving trams, projecting movie films, heating ovens and whirling around the tens of thousands of electric motors. Some of its grandest victories lay in the future, but the promise was there, nevertheless. The Electric Revolution presented men with regiments of early gadgets and devices which extended him—from ovens to wireless sets, tramcars to electric chairs. Human potency was amplified countless times. Since electricity is morally neutral, only a precarious future will show whether, in fact, Swan, Faraday, Ferranti etc really did present us with a power about which it could properly be said, *Fiat Lux*.

NOTES

1 See Chapter 5.

2 Lecture in Bristol, 8 November 1878.

3 Dropped in the early 1960s when Ediswan was lost in AEI, and now, therefore, part of GEC.

4 A portmanteau word from *Osmium* and *Wolfram*, constituents of the filament. A magnificent collection of early electric light bulbs can be seen at the Museum of Science and Industry, Newcastle upon Tyne.

5 Private information, kindly retailed by A. James, Esq.

6 Cmd 622 of 1920.

7 The vacuum cleaner was invented by H. C. Booth in 1901—his very early 'Goblin' machine worked by a mechanically induced vacuum, however. The first electrically operated ones were rather big, and were used for cleaning offices; they were often American, although the French had them in 1903. The 'Hoover' first appeared in 1908, and erected its first British factory in the electric suburb of Perivale, Middlesex, 1932.

8 Not at all safe, in fact. The only safe rule is to have no electric points of any kind in the bathroom, a rule later enforced by distributors.

9 There were more telephones being used in the British forces fighting in the Afghan War (1879) than in the City of London.

10 *Life of W. T. Stead*, by F. Whyte Vol 2; and *British Sea Power 1884–1905*, by Arthur C. Marder. See Appendix VII, part (2).

3, *Preparation and Reception*

ALTHOUGH electricity potentially offered a superior means of lighting and power, and it did, indeed, make great strides in actualizing its promise by 1930, it is fruitful to ask: why was the revolution no quicker? Why was it more rapid in Germany and the USA? By 1925 British consumption of electricity (*per capita*) was only two-thirds of Germany's figure, and two-fifths of that of the USA, which was a very marked inferiority, even making allowances for certain natural and geographical advantages of Britain's rivals. In the 1930s British consumption *per capita* grew fastest in the world, making up for lost time. The question remains: why was the time lost? It is clear that although the Electric Revolution was rapid and startling to its contemporaries, it could have been faster but for certain well-defined obstacles in its path. The setting which awaited the Electric Revolution was not friendly, and for all the technical brilliance of the electrical engineers and scientists, the electrical 'movement' had a tough and uphill struggle before it could relax, accepted and established.

Electricity's first battle was with rival forms of power, basically direct power from coal, or indirect via steam and gas. During the revolution, electricity never beat its enemies, although it had a strongly upsetting effect on them, and cut them down to size. In one sense, it captured steam power for itself, and with the demise of the steam locomotive on British Railways (1967), electricity generation remained the sole large user of steam power on land. In a similar way, electricity has been the saviour of the coal industry; 70 per cent of electricity comes from coal-fired power stations (1970), hence the hostility of the National Coal Board towards nuclear power stations which will seduce its single best customer.

It was all very different at the turn of the century. There were then about one thousand gas undertakings in Great Britain; they were not always popular or efficient, yet they were generally competent and well entrenched. Not a few belonged to town councils, which therefore eyed electricity as an unwelcome upstart that might undermine the corporation gasworks. At that time gas was largely used for lighting. When electricity proved its superiority in this field, gas went over to cooking and heating where it could and did offer much more effective opposition. Even as late at 1920 the value of gas was statutorily reckoned in 'candlepower', henceforward it was to be 'therms': the cautious Board of Trade's recognition of the great shift in the gas industry. The gas undertakers fought electricity hard, both by propaganda (see p 132) and by politics. One of their more effective weapons was to petition Parliament during the committee stage of an electricity bill, and to ask for a 'Bermondsey Clause' to be inserted.[1] This usually demanded that after the first year of operation of a municipal electricity undertaking, its receipts must balance its expenditure. Ostensibly this was to prevent the undertaking being a burden on the rates, and giving (in effect) subsidized power to consumers. In fact it was a millstone around many a young undertaking, which dared not embark on ambitious expansion plans for fear of running into debt. In this way the more active and aggressive gas companies choked some promising early power schemes.

Cheap and abundant coal was the foundation not only of the gas industry, but of domestic heating, railway transport and much of industry. It was the age of 'King Coal', of 'carboniferous capitalism',[2] when, as the economist Stanley Jevons had it, 'coal commands this age'. In industry, the steam engine in its various forms, or the rather more efficient gas engine, provided ample and reliable power. They were not so flexible as later electric motors, but they were established and trusted. Also, Great Britain as the oldest industrial power, was suffering not a little from managerial inertia. American and German business men were far more willing to give electric power a try than were their British opposites. By 1907, according to a government census of production,[3] one quarter of Great Britain's 9,650 million hp was going to generate electricity, but most of that was for lighting, not power. In 1901, one of HM Inspectors of Factories had noted: 'in the age of steam this country led the way, whereas in the age of electricity we seem to follow America and other countries', although another Factory Report of the time (1903) showed that electric motors were increasingly popular in small factories and workshops. By 1913 it was claimed in another report that 63 per cent of the workshops and factories in St Pancras, and rather more of those in neighbouring Holborn, were electrically powered; yet all of them were only modest in size. There were some really large users in Birmingham, Manchester, and of course, Tyneside, but all of them together could not lift the British electrical performance from its place in the international league, below the USA and Germany (see p 19), and on some statistical bases (for example, units consumed *per capita* per year), Switzerland and Norway. The proselytizing of Merz, Ferranti, Crompton and others helped electricity to infiltrate into shipyards and engineering; in some towns, councils eased the load factor by building electric tramways. However, it was the Great War that really showed what electricity could do as an industrial energy source, quite late on in the Electric Revolution.

In spite of the electrification of over two thousand miles of tramway, Great Britain had nothing to compare with the American 'Interurban', which did so much to spread the use of electricity in the USA. The interurbans were, in a sense, electric light railways that filled up the gaps in the 'heavy' steam railway reticule. At their height, about 1914, there were fifteen thousand miles of them, and this represented an investment of some thousand million dollars. They were by no means mere 'tramways' —their cars could run at high speeds, and they could run sleeping cars, diners and various other refinements. The interurban empire collapsed very quickly when popular motor cars appeared in large numbers; there were only three thousand miles of them in 1940, two hundred miles in 1960. However, they had been a strong spearhead for electric development, and many of them had been owned by utility interests, often large ones like Samuel Insull's financially disreputable combine, or the American Waterworks and Electric Company. Many of these utility systems had to shed their interurbans in the New Deal era after 1933, and without this valuable protection the remnants of the system perished swiftly.[4] Great Britain had nothing approaching this electricity-and-traction combination, unless one counts the British Electric Traction Company's rather disjointed empire (see p 42) nor, in view of hostile legislation could it ever have supported the necessary relationship.

Electricity men were wont to blame the clumsy laws relating to electricity supply for the slow development of their industry. By doing so, they played down the role of coal, gas, and steam power. Nevertheless, their argument had some validity, and it is amplified in Chapter 4. Suffice it to say here that the two basic Electric Lighting Acts (1882, 1888) enshrined a regrettable liberal dogma—universal competition—in an industry where it was unrealistic, unsuitable, and even dangerous. Electric power generation was not like the weaving of calicoes, or the rolling of steel —in which processes competition was sensible enough and had enjoyed great success. Electricity required a regional, or national framework in order to reach its full potential. Both Acts threatened private undertakers with municipal takeover, and so dampened enterprise, anyway. Theoretically no undertaking had a monopoly. In the best *laissez-faire* tradition the Board of Trade could, and occasionally did, back two franchises for the same area: one for a DC undertaker, the other for an AC one! Such ludicrous competition was fortunately rare,[5] but the patchwork quilt of supply by companies and local government did develop. On to this quilt was imposed the 'battle of the systems' (DC *versus* AC) so that neighbouring systems generated and distributed different types of power, thereby compounding the inherent nonsense in the 1882–8 legislation. Thus, even if the law had allowed neighbouring undertakings to co-operate (which it did not allow until 1909) it would often have been technically difficult, or impossible to do so. A few local authorities acted jointly, as one undertaking, from the start (like the Stalybridge, Hyde, Mossley and Dukinfield Tramways and Electricity Board—SHMD—see figure 35), but this sort of action was very rare, and fierce parish-pumpism was the rule. The big municipalities like Liverpool, Manchester and Birmingham, with city tramway networks and large numbers of consumers spread over a broad territory, could enjoy a more even load pattern, but smaller town undertakings did not. The McGowan Report (1936) put the matter succinctly: consumption per user, distribution and capital efficiency rise with the size of the undertaking, while costs fall.

Because of all these things, electricity was rather expensive in many areas. Eight pence (3½p) per unit was the usual maximum allowed by the Board of Trade, and this amount was often charged: that would be roughly five shillings (25p) in the early 1970s, whereas a typical charge (1970) was in fact about a penny halfpenny (1p) per unit at the same current price—eloquent testimony to the economies of scale possible in electric power generation, and in an age of inflation at that. As electric power became cheaper in later years, when organization was rationalized, so it ever tapped new strata of consumers. In turn, as real wages have risen, so more has been spent out of increased purchasing power on electric gadgetry and, therefore, electric power.

For this reason, the reorganization of electricity supply from 1926 onwards constituted a very real social revolution.[6]

The Weir Report (1925) criticized the electricity industry for being engineer-dominated, and marked this as a cause of its backwardness. For their part, the engineers often criticized the managers of undertakings for lacking real faith in the efficacy of electric power. Weir thought that the engineers were 'too engrossed in the race for lower generating costs', and since many managers were engineers anyway, high pressure salesmanship was not common. In fact, Weir spoke just as the 'All-Electric' movement was getting under way. It may have been stimulated by his strictures, but probably Weir's statement was altogether too *a priori*. Many manager-engineers were busy men, fully conscious of the need to improve their load factor: moreover, they knew the only sure way of doing this was to 'build load'—a favourite phrase of the time. Ferranti had said this back in 1910, and it had been quite common policy both before and after that.

The Great War had given impetus to the use of electric motors in industry,[7] and this was later backed up with publicity and attractive tariffs. To some extent the latter merely added to an already complicated tariff structure that some writers have seen as a barrier in itself to the use of electric power. One very real attraction to customers was the laying of mains in anticipation of demand, although local authorities were often reluctant to commit funds for this purpose (*pace* the Bermondsey Clause in some cases); they were, however, most strongly urged to do so by that doyen of municipal enterprise, the late Herbert Morrison, when he was Minister of Transport (1931) and therefore responsible for such matters.[8]

The Electricity Supply Act (1926) which followed on the Weir Report was launched amidst talk of 'cheap and abundant electricity for all'. It set up a national grid which was to connect a few efficient power stations, and scrap the plethora of local dwarf-stations with low load factors. A combination of this policy with continued engineering improvements began to achieve results. All-Electric propaganda and deflation made the process yet easier. By law, most undertakings had to pass back profits to the public after a given level, in the form of lower charges; municipal systems could use power profits to support the rates. At the laying of the foundation stone of the great Battersea power station (1931) the chairman of the London Power Co stated that £2,500,000 had been handed back to the company's consumers from 1925–30 because of greater generating efficiency, and the Electricity Commissioner's returns (1930) show that 2·8 per cent of the municipal operating surplus went to relieve the rates.[9] Such were among the early results of rationalization and good engineering practice. However, in the 'electric dawn' engineering was not sufficiently advanced, and politics were too dogmatic for such things. Even the most vigorous undertakings were given cramped areas in which to operate; undertakings which, by the laws of economic nature, ought to have been allowed to expand and absorb their weaker rivals and neighbours. After 1926 some quite impressive amalgamations did take place,[10] but even so, the McGowan Report (1936) thought that all undertakings selling less than ten million units per annum ought to be scrapped and compulsorily merged into bigger systems; and J. R. Beard[11] put the figure at twenty million units in 1940.

By 1906 local authorities had invested twenty-seven million pounds in electricity supply, and the output of this sector exceeded that of the private one: by 1914 the 327 local government undertakings had an output of 705MW, and of the 230 companies, 430MW.[12] Companies tended to serve small towns, or the sparsely populated areas between them. There was another reason, in addition to stated burdens that weighed on the private sector, as to why its progress had been so slow. Between 1870–1914 there was a burst of overseas investment from Great Britain. From one thousand million pounds so invested (1870) the figure rose to four thousand million by 1914. Much of this capital, which went to build railway and utility systems in Canada, Brazil or Argentina might have gone into home enterprise, notably electricity supply and engineering. The early supply companies, and particularly the larger scale power companies (see p 38) found raising capital hard. A single failure like that of the South Wales Power Co (in which eight million pounds was lost) weakened the private sector yet further. In 1907, of twenty-one million authorized capital for electric power companies—to be distinguished from the more promising and established local distribution companies) only £3,700,000 could be raised.

In electrical engineering the situation was no better. The stodgy and complacent nature of much industrial and commercial enterprise starved this industry of capital also. Capital could be found for textile mills, or the overseas Golconda, but not for this new industry at home. The field was open for foreign enterprise. American electrical engineering companies moved in where Britons apparently feared to; the American GEC[13] set up a subsidiary: British Thomson-Houston, and similarly British Westinghouse was established by an American parent. German interests set up Siemens and British Schuckert; AEG (Allgemeine Elektrizitäts Gesellschaft) had an offshoot also. Later, these became British (and most are in the GEC combine now), but their early history is an indictment of Edwardian enterprise, or rather lack of it. Similarly, the three main London 'tube' railways were foreign financed; they cost eighteen million pounds, most of it American money.

On the surface this seems a long tale of sluggish enterprise, missed opportunity and Gilbertian muddle. The early story of British electricity was all of this, but much more. It was a lusty infant which grew in spite of constricting conditions. The engineers, contractors and work force believed in the new power. Hostile attitudes about them merely sharpened their ardour and gave the 'cause' of electricity something of the flavour of a crusade. They cheered to a man at the meeting of the Institution of

Electrical Engineers as early as 1910, when the new President, S. Z. de Ferranti, made an uncompromising claim for nothing less than an *All-Electric Britain*. There was justice in the electric cause, whereas (it seemed) its opponents were technologically obsolescent vested interests, protected by mere political prestidigitation that would not conceal the truth for long. Thus, although electricity was born in an unkind climate, it was in no way daunted; it was the future, and it would work.

NOTES

1 So called because in 1902 the Borough of Bermondsey, asking for parliamentary powers to supply Rotherhithe with electricity, fell foul of the South Metropolitan Gas Co, which not only supplied Rotherhithe with gas, but was also a substantial local ratepayer. The South Metropolitan Gas Co objected to thus subsidizing a rival, and obstructed the Bermondsey bill in the House of Lords committee investigating the subject; it successfully persuaded the committee to insert a clause requiring Bermondsey's electricity undertaking to balance receipts and expenditure after one year. Attempts to cut corners by juggling with municipal accounts, i.e. for a borough to charge itself excessively for street lighting in order to 'balance' the electricity account were stalled in later bills. St Marylebone, Stoke Newington and Woolwich all received this treatment.

2 A. E. Smailes in *North England*.

3 *First Census of Production, Final Report,* which also revealed that three-quarters of electric generating equipment consisted of reciprocating engines, another mark of conservatism. By 1924, however, 90 per cent of electricity was generated by turbine sets—*Annual Report of the Electricity Commissioners* 1924.

4 See Hilton and Due : *The Electric Interurban Railways in America.*

5 It was found in Kensington (three systems), Bermondsey, Westminster, City of London, Greenwich and Camberwell, among others; it was generally confined to London (see *Marindin Report* p 56).

6 It was calculated at the time that price elasticity of demand for electricity began operating at the point when it fell to 2½d (1p) per unit.

7 See page 43.

8 Overall responsibility for supervising generation and distribution has rested with various government departments; they are : Board of Trade (1882); Ministry of Transport (1919); Ministry of Fuel and Power (1939); Ministry of Technology (1969) and Department of Trade and Industry (1970).

9 The profits of nationalized electricity, after 1948, were absorbed by (*a*) rapidly rising costs, and (*b*) a massive investment programme. These conditions did not exist in the inter-war period, of course.

10 Notably the NESCO combination, 1932 (see p 60).

11 *Journal* IEE, 1940.

12 In 1970 the chairman of the South Eastern EB estimated that 1,000 gigawatts (1GW = 1,000MW) would be required by AD 2020. Compare with 1·135GW in 1914!

13 As remarked elsewhere in this book the American GEC is no relation of its cognominal organization in Great Britain, although American GEC did make an attempt to take over British GEC at one time (1928–30). See Jones and Marriott, *Anatomy of a Merger*.

4, A Digest of the Laws—Politics and Power I

FROM 1882 until 1948 the various electricity laws and statutes moved from favouring parochial 'competition in the abstract' towards a comprehensive national network. These laws display interesting and significant social developments. They show the gradually growing power of government, in response to the need for rationalizing and humanizing a rapidly growing and complicated industrial civilization. As the power of the centralized state grew, so inversely that of particular interests, and of local government has declined. Within the state there has been a marked shift in influence from the legislature to the executive. The Electric Revolution shows three things most clearly: the political changes alluded to; the near-disastrous effects of doctrinal obstinacy in economic matters and the passing of an era in which undue deference (by later standards) was paid to interest groups outside of Parliament, notably the great municipalities. Partly, but not entirely, because of poorly framed laws, localism, whether private or public, was shown to be incapable of managing a national electrical system. Permissive schemes encouraging local co-operation failed miserably. In many ways this failure was pregnant with meaning for the future—it helped men to discover the need for 'Big Government', which would perform its task without over-sensitive inhibitions about local privileges and the sanctity of property. The gains made as a result of these changes were material and dramatic; probably the losses also, but they were of another kind.

Before describing the effects of these laws, it is necessary to describe them and the conditions which created them. Behind the first electricity law, the Electric Lighting Act 1882, lay the experience gained by gas and tramway legislation, notably the Tramways Act 1870. As so often, the wrong lessons were learned as men looked for guidance from the past to solve the difficulties of a strongly contrasting present. Lest this criticism be thought too facile, an important rider must be added. Given the state of engineering knowledge as at 1880, men legislated reasonably enough. Unfortunately, the state of such knowledge rapidly outstripped the competence of the political machine to adapt to it, with unfortunate consequences. The ability to generate and distribute high-tension AC over large distances at once made electric power totally unlike coal-gas, bound as it was then to very local systems. The Tramways Act had contained a clause permitting local authorities to compulsorily purchase tramways built in their areas after 21 years; a similar clause was inserted in the 1882 Act.[1] This Act was fashioned by a Liberal Government, which held to the tenets of Manchester School economics with some stubbornness. Particularly, it had a horror of monopoly in any form, hence its vision of granting many small, local franchises and forbidding any connivance between these enterprises. The President of the Board of Trade, who guided this Act in its bill stages, was Joseph Chamberlain, the sometime Lord Mayor of Birmingham, and as such the embodiment of vigorous local enterprise. The 1882 Act allowed local authorities not only to take out franchises themselves, but to effectively block private applications as well. Mutual co-operation between electricity undertakings was statutorily obstructed until 1909.[2] If business men had been left to their own devices, they would have sought economies of scale and amalgamated; as matters turned out they were not allowed to, and in any case many undertakings were municipal, and, as such, fiercely parochial. As late as 1919 a major statute foundered because of this; ten years before that the metropolitan boroughs of Islington and St Pancras thought that linking their systems in any way was 'an illusion'. In these ways, *laissez-faire* smacked of hypocrisy, or at least paradox: it forced businesses to remain separate, when if 'let alone' in the way that the Manchester School preached, they would have coalesced.

Government had intervened in the gas industry in order to prevent the possible chaos that threatened when gentlemen's agreements concerning territorial demarcation broke down. Many gas companies obtained private Acts, thus becoming 'statutory companies' or undertakers; as such they were granted a local monopoly in return for government fixed maxima on rates dividends or tariffs.This lesson was a fair one, and was later imposed on electricity companies, albeit in a modified way. Very few electricity companies proceeded without some form of state protection or *fiat*, for as such they were merely 'non-statutory' undertakers, and they relied purely on local good will to keep their position.[3] Whole sections of the 1882 Act repeated sections from gas Acts *verbatim*, and the impositions placed on tariffs by various maximum price clauses constituted an important step in utility control by the state.[4] Unfortunately, the gas industry was often highly unpopular with the consuming public, a fact which had baleful effects on the infant of electric power. It was partly because of this that municipal pressure groups insisted on the possibility that they could compulsorily purchase electricity undertakings after a given period of time. This concept was seen as a means of checking arrogant or irresponsible undertakers.

At first it seemed that municipal enterprise might be left the field of electricity generation to themselves. The Playfair Committee on local electricity supply (1879) clearly thought that electricity distribution was a municipal matter, but that if a local authority did not feel inclined to carry on the work, then a private company might be allowed to do so—on the clear understanding that it could be municipalized at a future date. The government did not immediately act on Playfair's recommendations, but with the dramatic inventions of Swan and Edison becoming widely known and popular it decided to produce the necessary statute to meet the probably wide demand for local electricity. The first draft of its bill mentioned compulsory purchase after seven years. This was extended in the committee stage to fourteen and finally to twenty-one years, which figure was duly enacted. This single matter is, in effect, the only important fact in the Electric Lighting Act of 1882. The twenty-one years' clause nullified all the rest of the Act, and caused it to be superseded after six very unsuccessful years. In the initial burst of optimism, some companies were formed (with a capital of fifteen million pounds), but none started operations. Of the sixty-nine Provisional Orders (see below) granted in 1883, not a single one was acted upon. 'The electric infant was strangled at birth' opined *The Statist*, prematurely, while the *Electrician* (1884) justly remarked that 'the fanatical dread of monopoly has resulted in there being no business either to monopolize or compete for'. *The Times*' original comments on the 1882 Act (when it was a bill) were prescient and correct: 'This bill will deprive the public of the benefits of electric lighting. The local authorities will not dare to embark on a comparatively unknown undertaking, and the private companies will refrain from doing so with those inequitable conditions of payment on compulsory purchase.'[5]

Friends of electricity, or progress, never ceased to attack the 1882 statute. In the House of Lords, Lords Thurlow, Rayleigh and Bury fought to extend the twenty-one year period and unleash some enterprise. Thurlow, speaking for the government, was sure that the electric enterprise waiting to be born would 'relieve unemployment' (the 'Great Depression' of 1873–95, rather modest by later standards, was pursuing its course), and he can largely take the credit for the successful passage of the Electric Lighting Act 1888. This brief statute was far more successful than its predecessor; it encouraged great enterprise, but because it still saw electricity as a local matter it provided a restricting framework that cramped and distorted the industry. Many of its ill-effects lasted until 1948, although some of the worst were modified after 1926. Its greatest achievement was to extend the period of initial enterprise from twenty-one to forty-two years, a more reasonable period in which to recoup investment before compulsory purchase.

Other clauses described the means by which an undertaking could get powers to start, and later the obligations which were imposed upon them when they had commenced the distribution of electricity. The simplest, but the rarest means of setting up business was by Licence, granted for up to seven years (and renewable) providing that the local authority agreed. A Licence did not need parliamentary consent, but very few were ever used. The more normal means was the Provisional Order (PO). A would-be undertaker served notice on any local authority likely to be affected by his plans. This gave local government an admirable opportunity for driving a hard bargain with the aspirant company, or for applying for a 'PO' itself—local authorities usually got preference over private companies in such applications. All this amounted, virtually, to local veto power; it had not been present in the 1882 Act, and had been added to the 1888 one by way of sugaring the 42-year clause pill, unpopular with local government. Assuming that no local objection was forthcoming, the Board of Trade granted a Provisional Order, so called because it did not have the full force of the law until it was confirmed by Parliament in a Provisional Order Confirmation Bill. There was, of course, another opportunity for obstructionists at the parliamentary stage, but the Board of Trade did not usually grant Provisional Orders unless they had a good chance of enactment. A Provisional Order did not grant monopoly powers, although outside London it implicitly did; and within two years of its being granted the undertaker had to start laying mains in specified streets. Consumers within fifty yards of a main had the right to demand connexion, although they had to pay the cost of being connected. There were various rate details in the Act; usually the rates were fixed in the Provisional Order, and they differed between undertakings, particularly within and without the metropolitan area.

The 1888 Act gave the Board of Trade fairly wide discretionary powers, and it could over-ride 'unreasonable' local objections to a Provisional Order. In this, the Act represented a big step forward in granting a department of state 'delegated powers', a matter which had caused some opposition in Parliament, jealous as always of its legislative monopoly. The fact that both municipal and private undertakings were set up in large numbers after 1888 does throw some doubt on the assertion that the 1882 Act was solely responsible for Great Britain's slow start as an electricity-using society. The twenty-one-year clause did not apply to local authorities, why then, did they not start operations? It is not really sufficient to claim that they 'waited upon' private enterprise as some kind of catspaw in a strange new world. Many Victorian cities were highly enterprising; some of the many other and very real obstructions to electrical enterprise have been already suggested (Chapter 3), and the slow start of electrical undertakings 1882–8 would seem to endorse them.

By 1900 there had been 639 Provisional Orders granted; a few had elapsed by disuse, 518 were in force (354 for local authorities) and a third of these were being materially acted upon. By the Electric Lighting Clauses Act 1899, the government standardized the type of Provisional Order granted, thereby imposing some uniformity on the conditions under which local electricity concerns worked. At the turn of the century the old 'Electric Lighting' law era had come to its zenith, and clear indications of its future problems were there for all to see. Electricity generation and distribution represented a new area for municipal enterprise, and some authorities like Cambridge, Portsmouth and Birmingham had taken full advantage of the opportunity. This was the great age of such municipal enterprise, a time when opulent town halls, corporation tramways, libraries, waterworks and baths manifested a lively local spirit, itself the product of non-conformist-radicalism and provincial pride, both equally determined to make up for the deficiencies of capitalism and central government. But there was an unattractive reverse side to all of this. There was the petty parochialism which obstructed co-operation between neighbour electricity and tramway systems; the 'dog-in-the-manger' attitude that led nearly 250 local authorities to sit on Provisional Orders without using them, so that at least nobody else could—an attitude often reinforced by fears for the municipal gasworks. In this way, merely to protect vested interests, or to keep a supposed capitalist bogey at bay, the people were deprived of electric light. This parochialism was intensified by different tramway gauges and electrical systems in some parts, and led directly to the tragi-comedy of that medley of voltages and frequencies that plagued Great Britain until the 1940s.

Large-scale AC transmission, made possible by the powerful turbo-alternator sets that were being constructed by 1900, was the single technological stroke that cut the ground from beneath the 1882–8 system, simply by destroying its technical rationale. When Joseph Chamberlain had introduced the 1882 Act, small-scale DC distribution from local stations was the only feasible system; but the talents of such men as Ferranti, Tesla and Parsons had changed that completely. However, as Lloyd George observed (see p 57) the question of the public supply of electricity is 'politics, not engineering'. The old Acts were backed by powerful interests that no government wished to tangle with: local government and sectors of business. Party discipline and executive power were not so well developed in 1900 as seventy years later, hence the reluctance of successive governments to grasp the nettle. Instead they began to shilly-shally and compromise after 1900, deferring first to one interest group, then another and in all supplying an unedifying spectacle of the state betraying its first duty—to protect the public interest against sectional ones. In this matter of failing to protect the common weal, the Liberals had the poorest record, victims of their own obsession about 'monopoly'. The Conservatives made a pragmatic and useful reform, albeit a compromise still, in 1926; Labour cut the Gordian knot in 1948. Yet it is not so much party-policy as the sheer lack of executive power held by the government when faced with obstructionism that is so remarkable to later observers. Before Hitler's War no government dared to attack the problem thoroughly; although many were advised to, none felt that it had the means.

'Science has overtaken statutes', observed Sylvanus Thompson in 1901; 'methods that were in vogue . . . in 1888 . . . are being superseded . . . larger areas of supply and fewer generating centres in an area are necessities of time . . . we are gradually passing to the use of multi-polar machines . . . ten years ago a distribution six miles from a generating station was looked upon askance . . . today it is recognized that the secret of economic working is to generate on a large scale and to distribute over a large area at an appropriate high voltage.'[6]

The first plan to generate from a big central station in the manner spoken about by Thompson, was Ferranti's master-plan for electrifying all of London from the Deptford power station of the London Electric Supply Corporation (1887). Both Deptford and the LESC were years ahead of their time technologically and functionally. Ferranti based his plan on what the Gas Light and Coke Co had done over ten years before, by building a mighty 'central station' at Beckton, on the Thames marshes, down-river, where colliers could come and supply a works that would feed gas to much of North and East London. Ferranti's station was to be south of the river, an integrated station transmitting 10kV AC all over London. His plan had to be severely modified after the 1888 Act, it remained nevertheless an example and an inspiration to electrical engineering.[7]

So long as Provisional Orders apportioned only individual towns to separate undertakers, long-range transmission was commercially pointless, although technically superior. In the late 1890's the idea grew that it might be possible to supply the large areas of no-man's land between urban areas, consisting mainly of small towns and villages, from large central stations. The promoters of such schemes hoped to include some of the larger and more affluent towns in their net works, but in view of the 1888 system, and the forces it manifested, this was to prove very difficult, and usually impossible. It was felt that the large areas in question might justify the establishment of power companies (as they came to be called, and as distinguished from 'distributors' who were the orthodox suppliers under the Provisional Order system), which could supply power to local undertakings requiring it, and perhaps to themselves if they could win some Provisional Orders in the area in question. In 1898 the General Power Distribution Co applied for a private Act of Parliament to supply the 210 square miles centred on Chesterfield; an area containing one million inhabitants. At once the city fathers of Sheffield and Nottingham detected a 'monopoly' threat, and demanded that their cities be excluded from such a plan which was 'a most perilous innovation, prejudicial to the interests of the great municipalities'. Deprived of such promising markets, the promoters of the General Power Distribution Co lost interest, and the proposal faded away.

However, the power company idea had many supporters, notably among the pragmatic Conservatives, and a parliamentary enquiry into the system (the Cross Committee 1898) reported favourably on the power company concept. It added, however, that there should be no general enabling Act (bearing in mind 1888) but that each case should be judged on its merits, and should have a full parliamentary enquiry *en route* to obtaining a private Act. In the next ten years some large power companies were launched with Parliament's approval: the Clyde Valley Electrical Power Co (735 sq miles); the Cleveland and Durham County Electric Power Co (820 sq miles); and the Yorkshire Electric Power Co (1,800 sq miles)—all dating from 1901, and all serving areas of considerable, if scattered, industry. The price paid by all power companies was their exclusion from the rich and populous urban areas which would have given them a better load factor, and enabled them to carry out more ambitious small-town and rural electrification by temporary cross-subsidizing. Even the most successful of these companies (the Newcastle upon Tyne Electric Supply Co or NESCO, see p 62) never included Darlington, Tynemouth and Sunderland in its territory, although it was fortunately based on Tyneside, having grown outwards from a city, very unusual for a power company. Most power companies lacked the firm foundation of NESCO, their areas looked impressive enough on the map, but their networks were often (and especially at first) very meagre. A power company supplied current to distributors, therefore a wise precaution was to take out Provisional Orders for oneself in order to have a guaranteed basis of consumers to start with. In this way the County of Durham Electric Power Distribution Co had ten Provisional Orders by 1906, including Durham City and Gateshead.[8] An additional help for power companies was the 'Kitson Clause'[9] which enabled them to supply bulk power anywhere in their area to certain enumerated customers like railways and docks, existing POs notwithstanding.

The Electric Lighting Act 1909 touched up this 'system' with further compromises. Although no local authority might compulsorily purchase a power company, a combination of authorities might do so after 42 years; given their touchy independence it seemed highly unlikely at the time that they would ever agree on terms when the hour struck. On the other hand, the 1909 Act at last allowed close co-operation between suppliers of all kinds.

The painful story of London's electricity supply (see p 55–7) led to similar, but less successful plans to ensure integration. In many ways the 1909 Act was the belated outcome of the Cross Committee, eleven years before. There had been three false starts with general power company statutes in Parliament since Cross's time. Cross, while rejecting the idea of an all-embracing enabling Act, did recognize the need for some minimum common pattern within which power companies might operate. There were such questions to settle as: did a power company have the right to compulsorily purchase land for a power station outside its bulk area? As it turned out, the 1909 Act did allow this, and a few other useful powers, but in general it was a mouse and did little to tackle the increasingly muddled pattern of electricity supply.

By 1914 the jig-saw pattern of supply and distribution was firmly settled. Big cities were usually served by their own corporation electricity department and its works, and they usually sponsored corporation tramways as a useful 'load builder' as well. Small towns were served by their own councils, or by private undertakers; all these suppliers, great and small, operated under Provisional Orders, except for a tiny handful of non-statutory undertakers. Between the Provisional Order areas there occasionally operated power companies, but these were sickly in their youth: by 1907 they had, together, an authorized capital of twenty-one million pounds, but only £3,700,000 of this was paid up. London was almost a law unto itself, and a microcosm of the greater muddle beyond Potter's Bar. This confusion was made worse by the wild variety of voltages, frequencies and the arbitrary adoption of DC by some undertakers, AC by others; all of this was the legacy of the pseudo-*laissez-faire* and 'competition' dogma of the Liberals, and the Board of Trade had to look helplessly on, punctiliously recognizing the variety of systems, and enforcing tariffs and safety standards within them.[10]

Two great difficulties faced any reformers of this muddle, such as Charles Merz.[11] The maze was jealously guarded by interests, which, while they might war amongst themselves, quickly united when faced by 'enemies' without. The maze had, in fact, a powerful legal basis that was going to prove very hard to crack. Because so many systems were technically incompatible, they remained small. Because of this, in turn, many of them suffered very low load factors; the average in 1907 was 14½ per cent, a low figure that also reflected the continuing tendency to use electricity for lighting only, with consequently sharp peak demands and low 'plateaux' in between. Some undertakers tried to build up a 'daylight load' of cooking, industrial motors etc, but the success of that campaign came after the Great War. Still, some progress in load building was most certainly made before 1914, mainly by offering favourable tariffs. Thus Kettering Urban District Council (1904) asked 4½d (2p) per unit for lighting, but 1½d–2d (1p) per unit for heating and power; Hastings asked 6d (2½p) per unit for lighting, 2d per unit for power, and 4½d per unit for cinemas, which demanded power coincident with the evening peak, and so were not let off so lightly (1909). Not all local authorities were so enterprising, and engineers as well as electrical publicists often criticized them for their lack of enterprise, which (it was suspected) often had something to do with the municipal gasworks, or (if privately owned) the shares held in it by the city fathers. On the other hand, there are some matters that can be submitted in defence of the undertakers. To create or satisfy quickly increasing demand would entail heavy investment, and probably initial losses. Even if no 'Bermondsey Clause' stood in the way of this, the fear of falling back on the rates was very real, and most ratepayers would have been up in arms if this had happened. To see ratepayers as stern progressives, eager to make sacrifices for the new energy, is quite inaccurate. Also, some authorities were pace setters in electrical enterprise; Birmingham and Hackney, for example, did much to popularize the 'All-Electric' ideas of Ferranti, especially after 1919.

The company areas were (with the exception of London) the lesser urban areas, or the wide open spaces between them. Even if amalgamations were allowed, they were rarely feasible because company districts were widely separated. One way around this was the 'Holding Company' which owned a number of small-town companies, or electricity works, and could practise economies of scale in a modest way. There was, for example, the Urban Electric Supply Co which owned electricity undertakings in some seventeen places, from Hawick to Camborne. Urban Electric was in turn owned by Edmundson's Electricity Corporation which owned thirty-two electric light companies in Great Britain.[12] Edmundson's was later a subsidiary (of the Greater London and Counties Trust; of the Utilities and Power Co of Chicago) and so a long web stretched out from Chicago to the Glossop Tramways or to the electric lights of Camberley, in Surrey. Few people knew all these details, but many talked of the electric 'combines' or 'groups', and not always with affection. At the same time these groups did impose some rationalization on the private sector of distribution and supply. For example, the Edmundson Group was closely associated with English Electric, whom it tended to use as its main supplier of electrical equipment. Similarly, the Greater London and Counties Trust dominated, by one means or another, large and continuous tracts of Cornwall, Wessex, the West Midlands and Northern Home Counties. Its 'subsidiaries and associates' included the Cornwall Electric Power Co, the South Wales Power Co, the Shropshire, Worcestershire and Staffordshire Electric Power Co, and many local distributors. Other financial interests included English Electric's Power and Traction Co, which, *inter alia*, backed the Falls of Clyde hydro-electric scheme, thus ensuring that the water-driven alternators at Stonebyres and Bonnington were of English Electric manufacture. There was the Power Securities Group, a name variously associated with British Thomson-Houston, Balfour Beatty, and BICC; British Westinghouse's Traction and Power (1901) was set up to finance railway electrification, with only modest success. A small handful of companies (usually local distributors) ran both gas and electricity undertakings, the only time that both of these public utilities enjoyed common ownership and direction of a really closely integrated type—as opposed to the strategic direction of a Minister of Power. Such included the Ascot District Gas and Electricity Co, the Guildford Gas Light and Coke Co, and undertakings at Salcombe, Bungay, Uckfield and seven other places (by 1948). The British Electric Traction Co (BET) owned about forty tramways, and nine very scattered electricity undertakings, at such places as Lewes, Sheerness, Penarth and Weston-super-Mare which otherwise had nothing at all in common.[13]

Complaints about these 'impersonal groups' go a long way to explaining the strength of municipalism in electrical matters. Although many companies were genuinely local and small, the existence of giants like BET really did excite fears of 'monopoly'. This was, of course, the era of the American 'Trust', and many radicals were determined to stop any such development in Great Britain—particularly in the field of public utilities. Thus it was noted with suspicion, that BET was associated with: Brush Engineering, Provincial Electric Lighting, 'Northmet' Electric Power Supply Co and hosts of tramways. It looked very much like an embryo-trust. Big towns applied for Provisional Orders partly to keep out of the clutches of these organizations, although state supervision would have made irresponsible rating or administration very difficult in Great Britain: the 1899 Act hedged in undertakers most carefully. Add to these suspicions the municipal pride of the era and one can see why so many towns preferred to generate their own power. It was a fit and proper object of municipal enterprise; it added to local amenities and it might ease the rates if properly run. Also, it might keep the town out of the darkling grip of the city financiers and capitalists, with their power syndicates and traction rings; as the London County Council put it when plans were afoot to rationalize London's transport: 'Hands off the People's Trams!'

The Great War showed how vital electricity had become, in spite of its lack of a national system. The output and investment of the industry leaped during the war:

Capacity of systems, 1914–18

Ownership	*Capacity 1914*	*Capacity 1918*
327 Municipal	705,000kW	1,490,000kW
230 Company	430,000kW	788,000kW

Ordnance factories had to be placed in safe and remote inland areas, where there was usually no local coal. The flexible nature of electric power supplied them with the energy they needed; thus was born the instant and mobile factory which later became a normal feature of regional development: 95 per cent of all new munitions factories were so powered. On the other hand, the deficiencies of the distribution system were thrown into hard relief, and as thoughts turned to the post-war economy, the 'groundswell of opinion' started to heave on the Electricity Question, as it came to be termed by some.

Late in the Great War, Great Britain was in a radical and reforming mood, stern and hopeful for the world after Armageddon. The slogan 'Homes fit for Heroes' is still well known, although usually used as a wry comment on the ephemeral nature of political fashions. The Prime Minister, Lloyd George, was a master of short-term manoeuvre, and quickly caught the feeling of the times; he launched a far ranging and elaborate policy of 'Reconstruction' whereby Britain would be reshaped along radical-liberal lines, and *inter alia*, a reformed electricity industry would supply the needs of the new society. During the war, some expert committees had profferred suggestions as to how to deal with the post-1888 chaos of electricity supply. *The Parsons Committee*

(1917–18) recognized that electricity was already a key industry, but badly divided by faulty legislation. Some sort of central body 'untramelled by the past' should co-ordinate it. *The Haldane Committee* (1916–18) thought that much wastage might be saved by 'central supervision' because the 'lowest possible cost of electricity is a condition of success for British Industry'. *The Williamson Committee* (1917) began to get to the heart of the matter by suggesting a body of Electricity Commissioners who would relieve various government departments (notably the Board of Trade) from their supervisory duties relating to electricity; the Commissioners would standardize supply and fashion a rational system of distribution from the labyrinth. Like the Haldane committee, Williamson looked quite far ahead, and at international comparative costs, and (bearing the nationally low load factor in mind) thought that 'success will depend upon the adoption of the most efficient methods of machinery'. Finally, the *Birchenough Committee* (1918) speaking for the more radical Ministry of Reconstruction, which was then at its most effervescent and aggressive, went the whole way and demanded a 'nationally owned supply network . . . not only a national system, but a single unified system, under state control'.[14]

Meanwhile, the interest groups had prepared for battle: a joint committee of the electrical manufacturers' BEAMA,[15] the power companies and the Institution of Electrical Engineers thought that gradualism was a far better solution, and this was widely supported outside the industry by many, who, although in favour of reform, leaned more to Burke than the Bolsheviks in their conception of the necessary speed of change. The Lloyd George coalition Government played the game quite skilfully, asking for more than it expected to get. Unfortunately, the post war radical phase evaporated more quickly than they had allowed for, and in the end surrender to the vested interests was all but complete. The first bill was a disaster; it had spoken of taking over all private companies at historic cost—in effect their original cost years before, unrelated to their 1919 values which were much higher owing to inflation during the war.[16] The bill was quickly withdrawn, but not before a great deal of trust and goodwill had been destroyed.

Two shifts in politics combined to produce the Electricity Supply Act 1919, the final outcome of 'Reconstruction', and a flabby compromise. All the undertakers, municipal and private, had been frightened by the Birchenough recommendations and were uniquely co-operative among each other, the municipal pressure group (IMEA)[17] joining with the power companies' IAEPC[17] both loudly telling the government that they would co-operate in future. By early 1919 Lloyd George sensed that Reconstruction was becoming played out, and he found more pressing problems to deal with: militant unionism and Ireland, to name but two. Also, his own political status made it necessary for him to conjure carefully with matters like nationalization. He was supported by a coalition of some Liberals and the Conservatives; the latter being in the majority. He had to steer carefully between their widely different concepts of the national interest, as a result he dropped the more radical schemes (like electricity nationalization), and it is hard to see what else he could have done, political realities being what they were. It is too facile to dismiss his action over electricity as mere duplicity. The 1919 Act was, nevertheless, a great missed opportunity; as so often in fickle democracy, the 'hem of fate' had not been grasped in time, and she did not return until 1948. A plan had remained in the draft second bill to set up District Electricity Boards with borrowing powers in order to enable them to purchase and co-ordinate power stations. A well-generalled press campaign sensationalized this clause, and drew lurid pictures of vast public debts, the proceeds of which were to be squandered on purchasing power stations. Once again, the government backed down, and the District Electricity Boards were dropped in the House of Lords stage of the bill. The emaciated remains of the bill set up the Electricity

Commissioners as state 'watchdog' of the industry, with limited but useful powers, and *permitted* the undertakers in various regions to combine if they wished into Joint Electricity Authorities (JEA's).

The Electricity Commissioners (a body which lasted until 1948) first consisted of Sir John Snell (chairman); W. W. Lackie (sometime chief engineer and manager of the Glasgow Corporation electricity system); A. Page (former general manager of the Clyde Valley Electrical Power Co); Sir Harry Haward (ex-comptroller of the London County Council) and H. Booth, of the Board of Trade. The heavy preponderance of a 'Clydeside brigade' raised a few eyebrows, but few disputed the choice of the eminent Snell as chairman; he was a lucid and competent Devonian engineer, with the requisite *gravitas* and reputation for a chief commissioner—he was currently (1919) chairing the Board of Trade Water Power Resources committee. The Electricity Commissioners' job was envisaged by the 1919 Act as 'promoting, regulating and supervising the supply of electricity' and they were empowered to carry out such researches and ask for such advice as they thought necessary. Their first task was to designate certain 'Electricity Districts', in each of which they were to hold an enquiry with a view to fashioning an official scheme that would render electricity supply cheap and rational in the area under review—by mutual agreement, however. Any powerful undertaker could vitiate such schemes by obstructive attitudes, or more usually by striking co-operative attitudes at an enquiry, but dragging heels thereafter. Some counsel, representing municipalities, adopted an openly hectoring attitude at public enquiries, to the disgust of the electrical press. Generally, however, the District schemes were sensible enough, if rather modest, and the positivist yearnings of radical minds to have some overall national plan for electricity supply was met as far as was politically acceptable; in other words, not very far. The Electricity Districts announced provisionally in 1920 were:

Lower Severn
Mid Lancashire
South East Lancashire
West Riding of Yorkshire (Aire and Calder)
Mersey and West Lancashire
North Wales and Chester
London and Home Counties
North West Midlands
South West Midlands
East Midlands
South Wales
North Lancashire and Cumberland

The second stage of the Electricity Commissioners' task was far more controversial; they were urged by the 1919 Act to attempt to get affected interests in each District (that is: authorized undertakers; county councils; local authorities; large consumers and 'others'), to set up a Joint Electricity Authority, usually called a JEA by its contemporaries. The JEA was a formal and collective body charged with the duty 'to provide or secure a cheap and abundant supply of electricity within its District'—it could become an authorized undertaker itself, and existing undertakers could be transferred into it with their various powers (including rights of compulsory purchase) by consent. The whole agreement was to be sealed by an Act of Parliament. Local government was invited to join in the JEA idea, but in the event very few authorities showed any interest. By 1948 there were four JEAs:

North Wales and South Cheshire (in effect the North Wales Power Co)
London and Home Counties
North West Midlands
West Midlands

The London and Home Counties JEA took over five years of wrangling, major rows and compromises before it was set up (1925), and it failed to achieve anything like the necessary rationalization of the supply problems in its large and confused

District. On the other hand, the West Midlands JEA was quite successful; it supplied power over a large and continuous portion of its District (mainly the Western part, including Shrewsbury), although the tough municipalities in the Eastern part of its District never allowed it to organize their area: Wolverhampton, Walsall and West Bromwich. In these ways the Electricity Commissioners made some limited progress in designating Districts, which represented modest objectives, but they made little progress in launching the more useful JEAs that represented an altogether more divisive concept: rational planning at the expense of local liberty. The four JEAs that were born served only a portion of their District as suppliers, in other parts of the Districts they were largely co-ordinating bodies, largely dedicated to preserving the *status quo*, and so it remained until nationalization.

The promised co-operation never appeared. A few Joint Electrical Authorities were born, North Wales's being the most comprehensive but otherwise the historic patchwork remained inviolate. 'The authorized undertakings have been given a grand opportunity—apparently they cannot take it,' lamented the *Electrical Review*, adding with a *cri de coeur* aimed at the power companies: 'is there no vision, no imagination among them?'; apparently not, for it seemed that 'the old parochial feeling has triumphed'. For a brief instant the government, dismayed by the failure of the 1919 system, contemplated taking out compulsive powers (1920) but the very elements which had caused the government to back down in the previous year were stronger now, and an economic depression was threatening, so the chance was let slip again. In 1922 a new Act permitted local authorities to lend to Joint Electricity Authorities so that they might build big central power stations, but this was opposed because it might make the Joint Electricity Authorities effective (which many municipalities did not want) and so little use was made of these powers.

In democracies it may take a long time to educate the public into accepting drastic but necessary measures. By 1925 it was clear, first that the parochial problem remained and that it was worse; secondly that the 1919 Act was a failure. In order to make the matter quite clear, the government set up a committee to prove the point (Weir Committee) and then prepared to do something radical at last. The Conservatives, who had effectively denied Lloyd George the legislation he wanted, set about producing it themselves—but six years later, under very different circumstances.

The Weir Report (1925) was not the only voice trying to educate public opinion and give the government the strength to embody the general will in the face of sectionalism. Lloyd George himself (now a back bencher) had some trenchant observations to make, for example (from his *Committee on Coal and Power*) in 1921: 'the only way out of industrial troubles and internal strain is to be found in the scientific production and utilizaton of power'. The Weir Committee, which was set up to review 'the national problem of the supply of energy' was not another mere report-manufacturer, or excuse for inaction. It was very quick in its work, clear and pungent in its style. The Weir Report said that five years of 'co-operation neither produced enough electricity, nor was it as cheap as it could have been'; there were too many small power stations, and too many systems. The answer was to have fewer, bigger stations, connected by a 'national gridiron' (always referred to as the 'national grid' subsequently) and to standardize, especially with regard to frequencies[18]—this latter point would cost £10,500,000. Whatever Weir said, political facts remained and the report and its enactment were both remarkable for their political sagacity. Rather than do the really necessary thing and scrap the whole muddle (an action which would have required a revolutionary situation such as Baldwin did not have, nor care to create) it was decided to superimpose the grid on the existing 'system' and to rely on gradual attrition of small producers to rationalize the system; above all there was to be a steady closing down of small and old stations. There was an element of 'divide and rule' in this; Birmingham had little fear that its new super-power station at Nechells would be a non-selected station, whereas some of the older power houses with reciprocating sets, belonging to remote District Councils, would be candidates for liquidation at an early stage.[19]

Much informed opinion was sceptical about the proposed changes, especially in view of the fact that statutes had hardly benefited the industry in the past. The *Electrical Review* expressed the line 'that long suffering industry has from the first laboured under a superfluity of parliamentary and bureaucratic control . . . the salvation of the industry is not to be brought about by the waving of weird "gridirons" '. The City of Birmingham electrical engineer, a Mr Chattock, spoke in pure *Manchesterstum* terms: 'self interest alone will lead to a concentration of generation'. Baldwin pressed on unperturbed, and the Electricity Supply Act 1926 duly set up a Central Electricity Board (CEB) with orders to build the grid and standardize supply; the Electricity Commissioners were to close archaic and inefficient stations if possible, and only to permit the erection of big and efficient new ones. Scepticism continued, and of the second stock issue for the CEB, 92 per cent remained on underwriters' hands. However, it set about its task vigorously and completed the first (132kV) grid by 1933 (see figure 14): later extensions brought it up to 5,000 miles by 1940. The spare generation capacity, which was wastefully high as late as 1929 (84 per cent) was reduced to a mere 16 per cent ten years later. A sum of £17,500,000 was spent on standardizing Great Britain to a system of 3-phase AC, at a frequency of 50 Hz,[20] and between 1930–6 the CEB saved the country some fifteen million pounds in new generating equipment by its ability to tap supplies of electricity from most big power stations, thereby saving on surplus stand-by machinery.

As stated, the pre-1926 system remained largely intact. The grid and the Electricity Commissioners gave the State the whip-hand over the networks, although a rather Gilbertian system of supply resulted. For political reasons (mainly the desire to mollify the established undertakers), the CEB was not allowed to own power stations: only the connecting grid. It bought all power generated by the undertakers, and then sold back to them whatever they required. Thus it was a type of wholesaler, a monopolist and monopsonist at once—this method of operations, however, did centralize planning and decision making (literally on an hour-by-hour basis) with one national authority, and the results were a spectacular improvement. The grid also played a big part in accelerating industrial change, by making power available in most populous regions of the United Kingdom, and in due course many of the emptier ones as well. Industry could, and did, gravitate to markets rather than coal bearing regions, now that coal was 'sent by wire'.

In 1930 the last electric frontier was being opened up. An embryo national system was at last in being, yet the chaos and system of 1888 was far from cleared away, and the Electricity Commissioners were very sparing of their power to hurry on the closure of old power stations. The National Government tried to speed matters on by 'doing a Weir' with the McGowan Committee (1936), but in spite of its recommendations about closing all generating-and-distributing systems selling less than ten million units per year, and a White Paper on the subject, nothing was done until 1948 when the Electricity Act 1947 nationalized the entire electricity supply industry[21] from the CEB (already a public corporation) to such small undertakings as the Melton Mowbray Electric Light Company. In the great co-ordination which followed, 562 undertakings were lost into the British Electricity Authority which set about cleaning the Augean Stables that had sixty years of accumulated confusion awaiting Lord Citrine and his men.

Let us, however, leave the old system at our appointed year: 1930. Appendix VI gives data for that year, and the McGowan report tells us that shortly after (1934) there were: the Central Electricity Board; the Joint Electricity Authorities (3); authorized undertakers (635); joint boards of local authorities (5); there were forty different voltages and nearly fifty generating sets were half-a-century old! Yet Sir Giles Gilbert Scott's Battersea power station led the world; it housed Europe's most powerful turbo-alternators, and was a milestone in environmental planning in that it emitted sulphur-free exhaust. The speed with which the CEB was standardizing frequencies astounded politicians and engineers; yet Will's *Law Relating to Electricity Supply* (6th edition, 1932), listed an intricate and vast network of tariffs which demonstrated, in some cases, Alice-in-Wonderland logic at its most bizarre.

Such was the system which somehow, in the traditional British way, muddled through: part envy of the world, part comic relief for it and almost wholly a tribute to the engineers, if not for the political interests they served. Yet even they put the matter right in the long run, and the story had a reasonably happy ending. Today the Central Electricity Generating Board can describe itself as 'the largest single integrated power system in the world'; and one feels that Charles Merz and S. Z. de Ferranti would have approved most heartily.

NOTES

1 Tramways Act 1870, section 43.

2 For this prohibition, see especially Electric Lighting Clauses Act 1899, section 3 of Schedule.

3 Hastings and Eastbourne did, in fact, get electricity supplies by proceeding under agreement with the local authority (but not with a Provisional Order) before 1888.

4 This increasing power of government was strongly resented by some,

see remarks of Colonel Makins in the debates on the 1882 Act, *Telegraphic Journal* for that year, *passim*.

5 No allowance was made for goodwill; as with the Tramways Act, undertakings could be bought at 'scrap prices'—a most foolish concession to localism.

6 Thompson's Presidential address to the Institution of Electrical Engineers.

7 Because Ferranti's scheme dated from before the time when Parsons turbines were generally available, he had to rely on immense reciprocating sets; after the 1888 modifications more modest engines were used. Deptford began transmitting in February, 1891.

8 The only two Provisional Orders that it used, in fact. The remaining ones were to ward off rivals, and remained disused mainly because BET (County of Durham's backer) was insolvent. Matters looked up when the County of Durham Electric Power Distribution Co was taken over in 1905—see Chapter 5.

9 Named after an industrialist MP who had the good sense to insert the first one.

10 *London Gazette*, 31 July 1906.

11 See Chapters 5 and 10; Merz tried to reform the London mess, supplied the necessary information for Haldane's report in the Great War and was largely behind the North East coast network, years ahead of its time in matters of integration.

12 And the Guernsey Undertaking, nationalized (or, at least purchased by the States of Guernsey) 1933. By 1937 the Edmundson Group controlled sixty-six suppliers.

13 Other traction and power groups were : National Electric Construction Ltd; Balfour Beatty; Provincial Tramways; British Thomson-Houston was a tramway owner as well as manufacturer. The Greater London and Counties Trust was set up by its American parent in 1926 and dissolved in 1937. There was also the Electric Supply Corporation (associated with Cromptons) which owned one tramway (Dumbarton) and 18 electricity undertakings.

14 *Parsons:* Report of the Board of Trade Departmental Committee on 'Position of Electrical Trades after the War', Cmd 9072 of 1918. *Haldane:* Strictly the *Merz* report for Haldane's committee 'Internal Report of the Coal Conservation subcommittee of the Reconstruction Committee of the Ministry of Reconstruction on "Electricity Supply in Great Britain"' Cmd 8800 of 1918. *Williamson:* BoT report on

Electricity Supply Cmd 9062 of 1918. *Birchenough:* Ministry of Reconstruction Advisory Council report of the Committee of Chairmen on Electricity Power Supplies, Cmd 93 of 1919.

15 British Electrical and Allied Manufacturers' Association.

16 One weapon to fight inflation is statutory unwillingness to allow, too easily, for repeated upward adjustment of charges.

17 Incorporated Municipal Electrical Association, and Incorporated Association of Electric Power Companies.

18 The main areas of exceptional (ie not 50 Hz) frequency being: Birmingham (25 Hz); Clydeside; South Wales and Deptford (25 Hz); North East (40 Hz). See Appendix VI, 3.

19 However, the means by which the Electricity Commissioners could close down a station that was too old or expensive in operation were made deliberately tortuous and difficult. These 'non-selected' stations continued to survive, although only 40 were left by 1944, at which date there were 141 selected, ie efficient base-load stations. One method the Central Electricity Board used to persuade undertakings to part with an old power station was to take over its capital charges, not strictly legal until 1936, when it was permitted by retrospective legislation. Although undertakers lost their power stations, they remained as distributors, buying power from the CEB.

20 By 1937 only 1·1 million of Great Britain's 9·3 million consumers used DC; by 1970 it was 400 out of 17 million.

21 A few non-statutory undertakers were omitted, and it was still in order to generate power for one's own use. Also omitted what was the (then) relatively new North of Scotland Hydro-Electric Board, dating from 1943 and having some other social purposes than purely generating electric power.

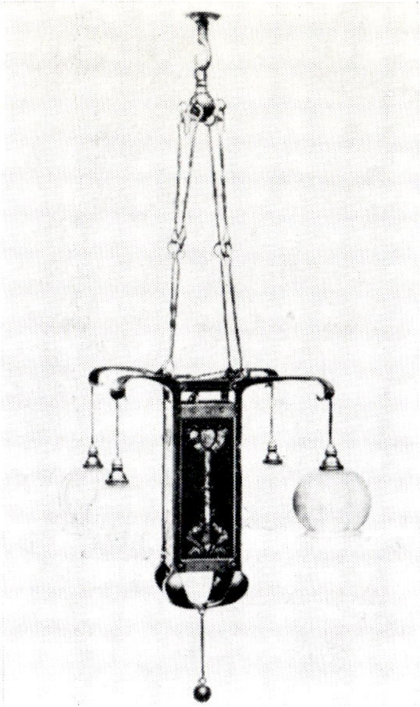

By 'Ediswan' in brass and copper

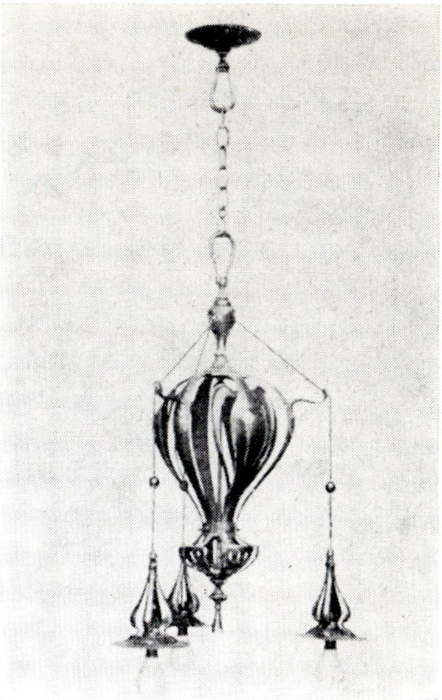

A cheaper line by 'Draulik'

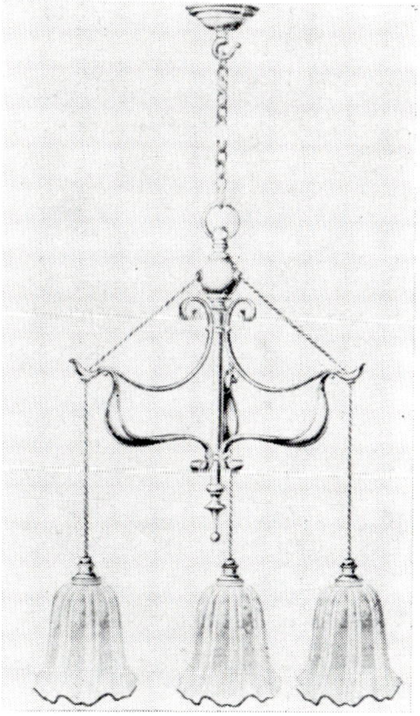

Early Electric! With what radiant hope
Men forged this many-branched electrolier,
Twisted the flex around the iron rope
And let the dazzling vacuum globes hang clear

Sir John Betjeman

Fine workmanship by 'Thursfield' of Birmingham 1905

FIG. 1. Edwardian electroliers: often hand-made, these *art nouveau* pieces graced homes which were able to afford electric light

General Electric Co's 'Dolphin' radiator showing special heating bulbs

Knife cleaner by Electromotors Ltd., supplied largely to hotels and restaurants

Electric coffee urn of 1905 by the British Prometheus Co of Kingston on Thames

Fig. 2. More Edwardian electric artefacts

FIG. 3. Chaos at Gosforth: a photo taken specially for a Parliamentary committee to demonstrate the inconvenience to passengers arising from the refusal of Newcastle upon Tyne (car to left of picture) allowing tramcars of the Tyneside Tramways and Tramroads Co (car to right) on their tracks—note break in conductor wire to prevent through running. A classic of electrical parochialism (1904). A compromise was later effected

Fig. 4. Lamp advertisements; (upper) of the early 1920's, and (lower) of the later 1920's. Although the lamp price ring (see page 14) kept prices at the same level, lamp manufacturers tried to create trade by lively advertising, note once-familiar house styles: owls, ballet dancers and the old Metrovick (Cosmos) monogram

FIG. 5. The General Electric Co's 'Osram' lamp works at Brook Green, Hammersmith in 1914. By that year the Osram works employed 1,500 people

FIG. 6. Electricity at war: the electrically-powered Territorial Army kitchen (November, 1914) which could supply 930 hot meals and 780 teas for three pounds thirteen shillings and fourpence (£3·67); power supplied by the Westminster Electricity Supply Corporation

FIG. 7. Economic nationalism: a GEC advertisement (1907) for arc lamp carbons, suggesting reliability, power and patriotism. Foreign competition was intense at the time; Witton (near Birmingham) has a long-standing connection with GEC

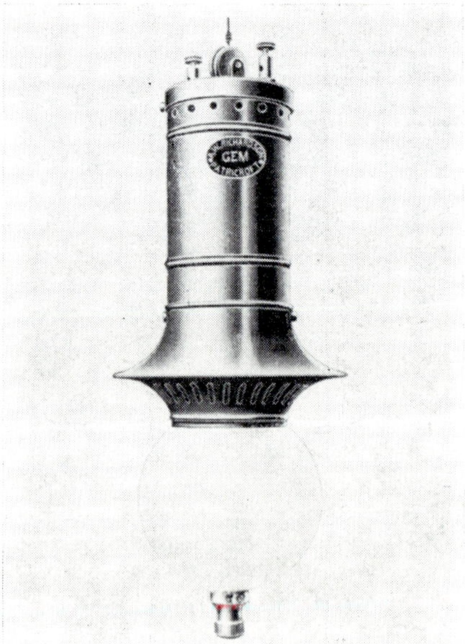

FIG. 8. An arc lamp: the carbon rods slid down from the long cylinder, and burnt behind the white glass bulb, this 'Gem' lamp was by A. W. Richardson of Patricroft; it worked on 110V for ninety hours before needing replenishment

FIG. 9. An early power house: the Gloucester Road station of the Brighton Electric Co (1887), with three Brush arc-lighting dynamos, driven by Fowler 'semi-portable' engines, clearly based on locomotive design, the far engine may have been a Hornsby

FIG. 10. The last grid tower: a momentous, and woefully underplayed occasion (5 September, 1933). The last 'pylon' of the original national grid being erected at Breamore, near Fordingbridge, Hants. At a celebration to mark the occasion, J. W. Beauchamp (Area Manager of the Central Electricity Board, see page 136) said: 'in the next ten to twenty years the home life and outlook of the housewife (will) be changed . . .'; it was, and far more besides. One of the great symbolic facts of British history has been largely forgotten: let this tower be gilded as a fitting obelisk to the Electric Revolution!

FIG. 11. A small undertaking: perhaps the smallest, Balderton (near Newark, Notts.) in 1914: the highway is the Great North Road! The premier road of Great Britain was the reason for Balderton's electricity—the council bought power from the nearby factory of Messrs. Jas. Simpson (at one penny per unit, say ½p.) to light twenty-four 100W bulbs, mainly 220V Mazdas, mounted on the creosoted poles to the right of the road

FIG. 12. Another local power station; the Basingstoke Corporation Electricity Works, built 1912–14. The plant cost £13,000 and was unusual in two ways: it employed diesel engines (two 150 hp Morley-Güldner sets) and generated DC, not in itself a rarity then, but uncommon for new stations by that date. An advantage of DC can be seen in the 300 amp Hart batteries, (lower illustration) which could store power and ease peak demand

FIG. 13. The local power station: in this case, of Kettering Urban District Council (1904); built for £43,000 and fitted with Willans & Robinson vertical reciprocating engines, driving Parker six-pole dynamos

FIG. 13a. Four Richardsons Westgarth 1,500 kW sets at the Weardale waste heat station, Spennymoor, Co. Durham, (1907). The steam for these sets was generated by waste heat and exhaust from the nearby ironworks, a practice once common in the North East

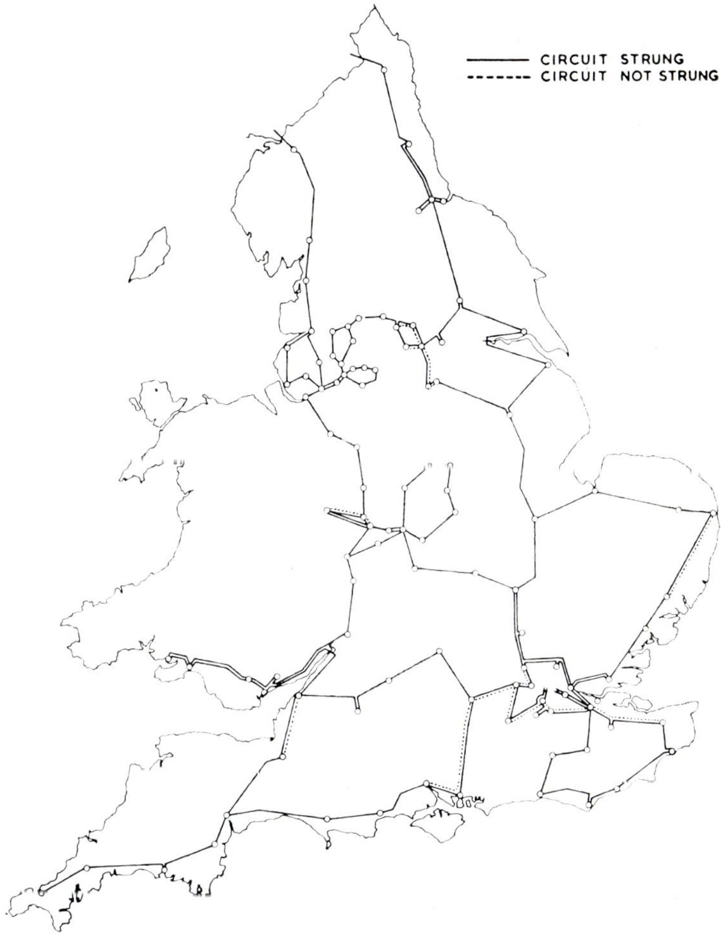

Fig. 14. The first (132kV) national grid in England and Wales (1930)—three years to go to completion

FIG. 15. Tidal and hydro-electric power: one of the many proposed Severn Barrages, this one was much discussed in the mid-1920's; others have been suggested in 1918, 1933 and 1945

FIG. 16. The Falls of Clyde scheme: Stonebyres hydro-electric station, erected by the Lanarkshire Hydro-Electric Power Co 1924–7, ordered to be plain and unpretentious by Parliament, to be 'more in keeping with the grandeur of the surroundings'; already people were conscious of a shrinking environment. Within the power house were two 2,500kW alternators, by English Electric

FIG. 17. Ardnacrusha hydro-electric station (1925–9), part of the Shannon scheme; mainly the work of Siemens-Schuckert. When it opened its 85MW capacity was sometimes sufficient for the whole Irish Free State

FIG. 18. Engineer's gothic: the Chester hydro-electric station on the River Dee (1911–13), remarkable both for its low head of water (six feet yearly average) and its being an English, as opposed to British, or Celtic, installation. It was built of concrete faced with sandstone and sensibly blended with its environs; the whole scheme was masterminded and carried out by the talented S. E. Britton, for many years Chester's electrical engineer, and a pupil of Colonel Crompton

Fig. 19. Michael Faraday, 1791–1867

FIG. 20. Sir Joseph Wilson Swan, 1828–1914

FIG. 21. Sebastian Ziani de Ferranti, 1864–1930

Fig. 22. Charles Hesterman Merz, 1874–1940

Fig. 23. James Staats Forbes, 1823–1904

FIG. 24. The all-electric factory: Arcadia House, built for Messrs Carreras, cigarette manufacturers, in the Egyptian style (1930)—it housed four hundred electric motors, powered by St Pancras Council Electricity Works. Note London County Council electric tram gliding by; the new electric world of the affluent South

Fig. 25. Summer day in Metroland: one of the Metropolitan Railway's neat electric locomotives (Metropolitan-Vickers, vintage 1922) shimmering through Middlesex. The 'Met' served the new electric suburbs: Wembley, Kenton, Pinner, Northwood, many of them also served by the vigorous 'Northmet' power company. Two of the electric locomotives bore names high in the roll of honour of electrical history *Michael Faraday*, and *Wembley 1924*—the exhibition that had done so much to popularise the new power

Fig. 26. A picture from the *General Electric Review* at the time of the first national grid (1927–33) expressing the hopes and contrasts arising from it. The picture was captioned with Tennyson's lines from *Morte d'Arthur*: The old order changeth, yielding place to the new

FIG. 27. Britain's first central control room, at Carville power station (1904) of the Newcastle upon Tyne Electric Supply Co. The station, and its central control for a large regional grid, was the brainchild of Charles Merz

FIG. 28. Tramway baroque: the fine wrought-ironwork around the bracket arms of the posts carrying the conductor wires for the Isle of Thanet Light Railways (Electric) Ltd.,—later the Isle of Thanet Electric Supply Co, which worked an integrated power-and-traction network in North East Kent. This picture was taken at Ramsgate harbour: note sailing vessels' masts intermingling with the new electric wonders, *circa* 1905

FIG. 29. Greenwich power station: built by the London County Council (1906) to power its electric tramways, this station had brick curtain walls and a steel frame. It was designed by W. E. Riley, FRIBA, who was obliged to shorten the southern chimneys (to the right) in deference to the wishes of Greenwich Observatory, further to the right and out of sight. For the same reasons, LCC trams had special earthing in the area—effected by erecting special return wires

FIG. 30. Greenwich power station within: some of the mammoth slow-speed alternators (94 rpm) coupled to reciprocating engines working on the 'Manhattan' principle (cylinders at right angles); a mode favoured in the USA. Slow speed alternators were more common in America and Germany, and were rendered obsolete by the steam turbine, a more compact machine. Vaguely reminiscent of Fritz Lang's *Metropolis*, these engines powered sixty-eight miles of tramway; each set generated 3,400kW

FIG. 31. A local station: Chichester diesel power house (1908), a type used where coal was relatively expensive, (see figure 12). Here, the two 3-cylinder Mirrlees diesels of the Chichester Electric Light and Power Co supplied two thousand lamps and some power consumers. Note iron 'pot' silencers

FIG. 31a. A large central station; the Edinburgh Corporation plant at Portobello dating from 1923, with later additions in the functional style

FIG. 32. Laying the mains: a big turn-out for an historic occasion at Torquay (1898), where the Corporation had ordered British Insulated Wire Co mains, laid variously in Doulton conduits, or wooden troughs

Fig. 33. Electric grandeur: a striking picture of Battersea power station, Sir Giles Gilbert Scott's masterpiece erected for the London Power Company. It was planned in 1927, and the foundation stone was laid on St George's Day, 1931—to commemorate the centenary of Faraday's *annus mirabilis*. By 1933 it housed Europe's largest alternator. Note the exhaust: cleaned and sulphur-free

FIG. 34. Early turbines: a switch gallery and two 1,500 kW Parsons turbo-alternators in Sheffield Corporation's Neepsend power station, *circa* 1912

FIG. 35. 'SHMD': the Stalybridge station of the magnificently named Stalybridge, Hyde, Mossley and Dukinfield Tramways and Electricity Board. A neat Edwardian structure in the valley of the River Tame

FIG. 36. Private power: a giant gas-engine driving an alternator at the Bargoed works of the Powell Duffryn Steam Coal Co—this one worked off their coke-oven gas, and drove 170 electric motors, hoists etc. Like many large gas-engines it was German, in this case by Maschinenfabrik Augsburg Nürnburg

FIG. 37. Early Electric: an LCC tramcar outside the 'Electric Theatre' at Clapham. Note conduit between the rails; a rarity in tramway practice, and an expensive but more aesthetic alternative to overhead conductors—early environmental planning

Fig. 38. Middle Electric: the New Victoria cinema (1930) eclectic architecture, electric light and power for all purposes

FIG. 39. Late Electric: the 500 MW generating set at Eggborough power station, Yorks, which broke all British records in the year ending 31 March 1971. Four such sets generated 7,711 million units between them, rather more than all of Britain's steam stations together in the year ending 31 March 1924. The sets are the property of the Central Electricity Generating Board, the 'largest single integrated power system in the world'—all a long way from the pre-Weir era

5, *North and South—Politics and Power II*

AT one time, before the Great War, it seemed that there might one day be a power station for every town. The Weir Report of 1925 noted that while twenty-eight large stations generated 50 per cent of Great Britain's electric power, at the other end of the scale 322 small ones generated 11 per cent. These small stations generally had a low load factor, and were consequently a burden on the national economy; before 1926 they were legally untouchable, and they vitiated the attempts of engineers like Merz who wanted to establish an efficient national network, however owned.

While there was some sense in Birmingham or Sheffield generating their own power, it was less obvious that Hebden Bridge Urban District Council or Rawtenstall Corporation needed to, unless one remembers that the 1888 Provisional Order system imposed this sort of pattern on the economy: if local enterprise was unwilling to set up a power network, there was little chance of anybody else doing so, certainly until the coming of the power companies. The menace to national efficiency only arose when localism refused to recognize or adapt to the turbine age. Local companies strengthened this pattern, and the fact that they were liable to compulsory purchase under the terms of the 1888 Act argued ill for the possibility of a rationalist solution in the future. A positivist planner looking through *Garcke's Manual*, or the *Electrical Trades Directory* for 1925 would have despaired at the proliferation of such enterprises as the Llandrindod Wells Electric Light and Power Co.

Until 1926, however, the worst confusion was to be found in the one area that ought, in view of its pretensions, to have given a national lead: London. The confusion of London's pattern of generation and distribution was really a function of its initial, and very early, enterprise. Electric power supply began in London rather ahead of the provinces, mainly because a large, rich and concentrated population was to be found in some of its districts. In the 1880s and '90s when this enterprise took place the power industry was a living laboratory, and the jumble of systems then available were built into London's electricity network. The Board of Trade had a mixed effect on all this: from 1889 it refused to issue large scale Licences (the seven-year simplified consent system, see p 33) but instead insisted on the normal Provisional Order process. A volume of applications was ready from many companies wishing to enjoy a slice of the London market, and the Board of Trade responded by commissioning a report from one of its engineering advisers, Major Marindin, RE. The Marindin Report (1889) favoured, *inter alia*, standardization of systems; big-area franchises and competition on a limited scale, ie AC and DC systems (working to the requisite standards) serving the same area, a peculiar and illogical blot on an otherwise sensible report. As a result there was some apportionment on a bigger scale than elsewhere in the kingdom, but many of Marindin's recommendations were ignored, to the detriment of London's social and economic future. By electrifying later, and not suffering the unique system of London local government, the provinces avoided this maze and displayed some internal cohesion.[1]

In 1917 it was reported[2] that 'in Greater London 70 authorities supply electricity to the public, and own some 70 generating stations, with 50 different types of system, 10 different frequencies and 20 different voltages'. Attempts to sort all this out ran into a political web: there were three powerful interests involved (apart from the customers, that is)—the companies, the metropolitan boroughs and the overall London County Council, all highly suspicious of each other. Progress might hang upon the result of a single election return, eg in 1907 when the Tory ('Municipal Reform Party') victory in the LCC elections ended a Socialist ('Progressive') attempt to bring about some overall order by statute. Merz had previously attempted to do the same thing privately, and in 1915 he tried again, backing the LCC in yet another attempt. Two bills were brought forward in 1915, the first fell foul of the metropolitan boroughs, the second of the companies; meanwhile London drifted on, acutely underpowered for a modern megalopolis.[3] After the 1925 Acts the London situation improved, and some really big amalgamations were allowed, particularly in the formation of the London Power Co which built the chief power station of the era, Battersea (London Electricity Supply Acts, 1925). Needless to say, the attempt to form a London and Home Counties Joint Electricity Authority took years, and it lacked bite when it was formed, in the same year as the reforming Acts. Such progress as was made came through the more forward-looking undertakers; Hackney was a paragon among the metropolitan boroughs, and 'Northmet' among the companies. Northmet[4] served many of the new Northern suburbs of London, and enjoyed great success in load-building, by propagating the All-Electric gospel. However, it was not until 1948 that London had uniformity imposed upon it at last, and received the kind of co-ordination that had become commonplace in provincial centres like Birmingham and Leeds for half-a-century. Significantly, it was in connexion with a proposed and abortive London electricity reform that Lloyd George remarked to the young Merz: 'not a matter of engineering, but of politics'.

In contrast to the kaleidoscopic London power map, that of the North East was almost blandly uniform. In many ways the North East coast stood in strong contrast to the rest of Great Britain in displaying a relatively integrated power network, operating over a large area: Northumberland, County Durham and Tyneside. It was not launched as such, an *a priori* rational system, but rather benefited from the vision and tenacity of engineer-entrepreneurs, notably the Merz family and their associates. Tyneside has many claims to be called the one true source of the electricity industry: here Swan made his discoveries on the incandescent lamp; Parsons perfected his reaction turbine; Charles Merz started 3-phase industrial power and erected the first real central power station with integrated control, in Great Britain.

At first it seemed that Tyneside might go the way of the rest of Great Britain, when two, later three companies commenced to supply power in the area: the Newcastle and District Electric Lighting Co (1889); the Newcastle upon Tyne Electric Supply Co (NESCO also 1889), and later, the Walker and Wallsend Union Gas Co (1901). The instrument that effected the great changes was NESCO coming to a demarcation agreement with its Newcastle rival and starting a career of constructive imperialism in the 'three rivers country'. Its chairman was Dr Theodore Merz (Charles Merz's father) and in later years Norbert Merz (Charles's brother) sat on the board of NESCO and many of its associates, as an interlocking director. Soon after the turn of the century there was a furious war in Parliament, the results of which settled the pattern of North East coast electricity supply, until and beyond 1948.

There were three contestants in the war: NESCO, the British Electric Traction-backed County of Durham Electric Power Distribution Co (which held some Provisional Orders south of the Tyne) and the 'Tyneside Power Company' backed by the District Co and some industrial interests. The prize was that part of Tyneside not yet covered by Provisional Orders, and perhaps the right to infiltrate or take over those parts that were covered. The battlefield was the parliamentary committee room at Westminster. The Newcastle upon Tyne Electric Supply Co won, partly because of its competent management, and partly because of the witnesses it could call upon. Its counsel (Mr Baggallay, QC) opened its case by chastising British Electric Traction with implications, suggesting that it might be able to run trams and light the streets by-the-by, but it could not rival NESCO at the really heavy work. One witness was Lord Kelvin, probably Great Britain's leading scientist and electrical academic. In his opinion NESCO could manage the North bank quite comfortably, and put mains under the Tyne in order to serve the South side with equal ease. At this point the war came to a sudden end; British Electric Traction called it a day, probably worried by Kelvin's assertion that NESCO could, technically at least, invade their territory. Certainly British Electric Traction could not reply in kind, for it was barely able to cope with traction and lighting in Gateshead.

By 1905 British Electric Traction was in financial difficulties, and sold its two County Durham concerns to NESCO, which was eager to buy this valuable nearby territory. The main reason for the NESCO anxiety to buy out the British Electric Traction interest was the great power station at Carville, opened in 1904. This station was designed by Charles Merz, and it was the first to house a central 'system control room'. Merz designed Carville on generous lines, and knew that it would require a big load to show off its efficiency to the best. Shortly after it opened, Merz had persuaded NESCO, the owner of Carville[5] to come to an arrangement with BET and supply them with much-needed power via a cable laid on the High Level Bridge between Newcastle and Gateshead. Thus a physical connexion between NESCO and County Durham existed before a corporate one, and the latter sealed an existing and convenient arrangement. The next move was in 1906 when Merz helped to reorganize the ailing Cleveland and Durham County Electric Power Co, which served the Durham coalfield and parts of Tees-side. The new firm was slightly renamed (Cleveland & Durham Electric Power Co) and it was arranged that it too, would take power from Carville by means of a 20kV cable laid under the Tyne (Kelvin's prophecy come true) to Hebburn. The Cleveland and Durham Co remained independent of NESCO until 1917, when NESCO bought its equity capital, thereby strengthening an already close relationship.

One final company remained, the Northern Counties Electric Supply Co (1900) which was based on Newcastle, but held a miscellaneous handful of Provisional Orders for various towns in the North East industrial region: Spennymoor, Morpeth, Ashington, etc. To Merz, its attraction was that it worked the Hebburn undertaking, just across the river from Carville. Thus, in 1906, as part of his Cleveland & Durham reforms, he arranged that the latter took over the Northern Counties Electricity Supply Co. Certain large towns never came into the fold: Sunderland and Darlington, for example, which went the municipal way. The ramshackle corporate organization of NESCO was considerably smoothed by interlocking directorates; a common headquarters; a regional 20kV grid and the common use of AC at 40 Hz, generated at Carville, Dunston, and on Tees-side. In 1932 NESCO obtained a private Act of Parliament, and the whole interlocking organization was welded into one company: the North Eastern Electric Supply Co (still NESCO) that lasted until 1948, broadly serving the same area as the post-1948 North Eastern Electricity Board, a tribute to the former organization which needed so little drastic reform.

Even in 1926 NESCO felt justified in congratulating itself thus: 'Indeed, the undertaking, as regards both the layout of the system and central station practice, has served as a model for similar undertakings both at home and abroad.' NESCO was a model organization; an indictment of confusion and localism elsewhere, and an inspiration for the future. Yet much of its achievement was fortuitous: the parlous state of British Electric Traction's finances; the support of Kelvin, the strange political geography of the Tyne valley, the family connexions of Merz,[6] but above all, the force and brilliance of Charles Merz himself. He saw the chances and exploited them with skill, and although he was a strict and peaceful Quaker, it would not be incorrect to call his work generalship of a high order.

To those who claimed that what had been done in the North East could be repeated elsewhere, there are two answers. First, the conditions on Tyneside were unique, and secondly, NESCO succeeded in spite of, not because of the electricity statutes—elsewhere the problem was often insuperable, and nationalisation was the only answer.[7]

NOTES

1 For example, in 1897 the Charing Cross Electricity Supply Co and St Pancras Vestry were among the nine DC suppliers; the Brompton and Kensington Electricity Supply Co and Islington Vestry (next door to St Pancras) among the seven AC ones. See R. H. Parsons, *op. cit.*

2 Report by the Electrical Trades Committee to the BoT, 1917.

3 Merz noted (1904) that 95 per cent of London's electricity was for lighting, with the consequent load problems.

4 North Metropolitan Electric Power Supply Co.

5 Carville station (1904–60) was owned by the Carville Site & Power Co, merged into NESCO, 1932.

6 Merz's uncle was the shipbuilder Wigham Richarson, also a shareholder in the Wallsend Gas undertaking. This built the pioneer 3-phase station at Neptune Bank (1901), and sold its power interest to NESCO in 1903; Carville adjoined Neptune Bank, and replaced it.

7 The full NESCO network before 1932 was: NESCO; County of Durham Electric Power Distribution Co; County of Durham Electric Power Supply Co (wholly owned by the former, and a power Co whereas the Distribution Co owned the Provisional Orders); the Cleveland & Durham Electric Power Co; the Northern Counties Electricity Supply Co; the Houghton-le-Spring & District Electric Lighting Co; the Tees Power Station Co; the Carville Site & Power Co. The chairman of most of these was R. P. Sloan, whom Merz's partner, William McLellan had brought with him from Siemens, to help build Neptune Bank at the turn of the century. A common director, Norbert Merz, also linked the NESCO system with the Waste Heat and Gas Electricity Generat-

ing Stations Ltd (independent of NESCO) that ran some small stations near coke and steel works; similarly he linked the Bankfoot Power Co, and the Priestman Power Co. NESCO also had arrangements with the Consett Iron Co (which generated its own power) and with the municipal 'islands' like Darlington, Sunderland, Tynemouth and Middlesbrough. Even in Newcastle, there were two moderate rivals, with their own areas—the 'District Company' and the City Corporation. These were important exceptions; nevertheless NESCO was unequalled in the extent and comprehension of its network.

Part Two

6, The Cradles of Power—Godalming, Brighton and Wallsend

THE early lighting systems mentioned in Chapter 1: Holborn Viaduct, the Savoy Theatre etc, were all tailor made for special and private purposes. None could claim to have laid on a public supply of electricity—that honour rests for all time with Godalming, in Surrey, which enjoyed (for a short time) the services of the world's first power station.[1] The *Telegraphic Journal* of 15 October 1881 reported that Godalming Town Council had accepted the tender of Messrs Calder and Barret 'to light up the streets with electric lamps', by using three Siemens arc lamps, and forty Swan incandescent ones, 'which will be placed in the lamps used by gaslights'. This was a sweet victory over gas, which (the promoters claimed) was nearly 20 per cent more expensive than electricity.

In fact, the street lighting contract was awarded to Siemens Bros, who also received orders to light a mill from their system; the contract was on a yearly basis, and power was turned on late in 1881. Thus, Godalming antedates all other public power supply systems, including Holborn Viaduct (second system); Pearl Street, New York and Brighton—all started in 1882. The Godalming scheme was progressive in another way: it was based on hydro-electric power, generated by the river Wey. It proved very popular, at first, for in September 1882 we read that 'a considerable extension will shortly take place', although the 'water motor' was evidently proving rather unsatisfactory. By 1884, however, Siemens were finding the scheme unprofitable, and their market research showed that few new customers were forthcoming. Generation stopped on 1 May 1884, and the gaslight returned to its old haunts.

Godalming's undertaking, although the first public electrification from a central power station, was abortive. The next honour —that of the longest history of continuous supply—goes to Brighton, which has received a public supply of electricity without break since 27 February 1882; it is claimed to be the longest record in the world. In late 1881, Robert Hammond staged a successful publicity exhibition in Brighton, for the Brush arc lighting system, and when it ended he was persuaded to give a week's trial (at twelve shillings—60p—per lamp per week) to shopkeepers dotted around a one and three-quarter-mile circuit, the system being powered by a 12-hp Robey steam engine in the yard of Reed's Iron Foundry, in Gloucester Road. Again, the experiment flourished, and Hammond was encouraged to form the Hammond Electric Light and Power Co, which generated on a more formal and permanent basis for all interested parties from February 1882. At first Hammond only supplied power from 'dusk to 11 pm daily', and for a total of sixteen arc lamps at that! From this modest start, the Brighton power system developed very quickly, a prodigy of success, typical of the buoyant growth of the electricity industry in virgin market conditions. In 1886 Brighton had eight miles of circuits, mainly carried in overhead mains of bare copper wire; and 1,000 incandescent lamps. In 1887 continuous round-the-clock service was given to the public at last. This was during the managership of the 'energetic and resourceful' Arthur Wright, who steered the Brighton undertaking from 1882–1905, and had to find his way to manage his charge with no previous body of experience to draw upon. His achievement is remarkable in retrospect, and he can take credit for: two-part tariffs, house meters for measuring the amount of current consumed, and a means of automatically maintaining voltage. Wright went over to AC in 1887 (1·8kV, reduced to

100V for domestic use) at the same time that Ferranti was advocating this current as the power of the future. Before introducing automatic voltage control, the company had employed a boy to perform this task; apparently he was apt to doze off, and Wright preferred to avoid further trouble by devising a controller of his own making, largely consisting of two electrodes that rose and fell in a tank of water, keeping voltage fairly constant: it is recorded that it did not vary more than 1 per cent from its proper value. The Brighton company changed its name and premises a number of times,[2] and was municipalized in 1894. In 1906 Brighton Corporation built an elephantine and functional power station at Southwick (to use seaborne coal) and, by way of a swansong at nationalization, it started to build the superb Brighton 'B' station next door, in the best 'Battersea-brick' style (1947); it was finished in five years[3] and now has a capacity of 324MW, the latter day inheritor of the duty started by Robert Hammond's Robey engine.[4]

The Godalming and Brighton schemes, like hundreds of other early ones, were regarded as purely lighting undertakings. The first major load-builder which gave power stations a higher load factor and therefore higher overall efficiency, was the electric tramway. Later, electric motors in industry were added to the load, although not in any big way until the Great War. The last laurel-leaf to be handed to the pioneers goes to Wallsend, where two power stations (Neptune Bank and Carville) were conceived and built as genuine central stations, supplying the needs of domestic and industrial consumers. By 1901 Tyneside had inaugurated the first heavy-industrial power network, at a time when most undertakings were compared in terms of how many 8-candlepower lamps were on circuit. The undertaking which pressed on with this progressive plan was, ironically, a gas company: the Walker and Wallsend Union Gas Co. Much of the corporate story of Tyneside power has been related in the previous chapter; there are, however, some important technical details to add in order to demonstrate how original the Wallsend plan was.

The gas company had on its board of directors Wigham Richardson, Charles Merz's uncle. Probably because of this connexion, Merz was invited to appear for the company before a parliamentary committee which was investigating the gas company's request that it be allowed to supply electricity. Merz was at that time engaged on establishing a tramway system in the city of Cork; he duly appeared at Westminster, made his case, and returned to Ireland. His evidence was masterly, the first of many such deliveries that he made before parliamentary committees. In time he was to be regarded as the expert witness without peer, in the meantime the gas company was so impressed that it asked Merz to become its chief electrical engineer. He declined, but agreed to be their consulting engineer, thereby establishing the other status for which he became famous. Merz brought with him William McLellan (late of Siemens) who had helped him in Cork, and shortly afterwards (1 January 1901) made him a partner. In this way the leading partnership of consulting engineers, Merz and McLellan was formed, and Wallsend was, indirectly, the cause of a remarkable hat-trick that was to shape the electricity industry thereafter: it brought Merz before Parliament for the first time, it set him up as a consultant, and it was the cause of the great partnership.

The gas company owned some land at Neptune Bank, on which Merz was to build the station. The partners gathered a team around them, many of whom were young men who were to rise to the highest ranks in the electricity industry, or in electrical engineering in later years. Merz decided that Neptune Bank was to generate and distribute three-phase AC, so that it could attract a wide market. In America, George Westinghouse was pressing the three-phase AC case, claiming quite justifiably that this system was superior to the Edison-GEC systems which were based on DC. The three-phase system was the invention of the remarkable Tesla, and one of its strongest recommendations was its ability to turn simple and reliable motors at constant speeds; ideal for industry. The main problem facing Merz, however, was to decide upon the frequency of his three-phase current. The normal AC frequencies were 25 Hz or 100 Hz, at the time. The former was admirable for motor and industrial purposes, but it flickered too much for public and domestic lighting. The latter gave a steady light, but was too high for the AC motors of the day. After prolonged research, Merz compromised at 40 Hz, which thereafter became the North Eastern standard.[5]

Neptune Bank was opened by Lord Kelvin (overcoming his anti-AC convictions for the occasion, see Chapter 7) on 18 June 1901. By the end of the year some large shipyards were connected to the gas company's electricity network, notably Armstrong Whitworths and the Wallsend Slipway and Engineering Co. By 1903, the gas company was working towards a friendly agreement with NESCO, the principal Newcastle supplier, of which Dr Theodore Merz was Chairman. The 'War of 1900' over supplying Tyneside had been won by NESCO, in alliance with the gas company, and during 1903 a more satisfactory situation was worked out whereby NESCO took over the gas company's electricity interest completely. Under Merz's guidance the gas company had been busy load building, and the 1903 settlement coincided with (and was partly occasioned by) a contract by which NESCO was to supply power to the North Eastern Railway's Tyneside suburban electrification. Neptune Bank was not able to cope with the gigantic load now waiting to be satisfied, even though Merz had two 1,500kW turbo-alternator sets added to its generating equipment. Instead, NESCO set about building an entirely new station, not far from Neptune Bank: Carville. Time was short, in view of the rapid progress of the North Eastern Railway's electrification. The Carville site was bought in January, 1903; the power station was erected and generating by 1 July 1904! Like Neptune Bank, Carville broke new ground. It contained a system control room, from which the whole NESCO network could be run. The idea came from America, but this was the first British installation of what has become universal practice since (see figure 27). Unlike the 'power house' appearance of Neptune Bank, Carville was large and stark, with steel chimney stacks and the now-familiar 'cathedral' layout of generating sets under a spacious roof. Its interior was not markedly different from the most modern practice; by 1904 the power station as we know it had arrived. Merz and McLellan jointly produced a paper on the subject at the time they were working on Carville: *Power Station Design*; it was delivered to the IEE and soon became the standard classic on the subject.

The idea of the central station was born at Godalming, 1881. Just under twenty-five years later it had fully matured, and was embodied, for all to see, in a genuinely modern plant. Carville lit, powered, heated and moved a large and settled urban area, and had within it that control room and desk, the successors of which have become nerve-centres of our civilization.

NOTES

1 Although the Rochester (New York) Electric Light Co offered hydro-electric power for arc lights at seventy-five cents per light per night, February 1880. The company was a main constituent in today's Rochester Gas & Electric Corporation, thus Rochester can claim continuous supply as old as Brighton's; the company was formed in February, 1880, although it is not clear when it started generating. It offered its services to all interested parties, whereas Godalming supplied public light for the whole town; like the Swan-Edison controversy, it is probably best to say the honours are evenly divided. See, *The R.G. & E. Story* by Arthur P. Kelly.

2 Brighton Electric Light Co 1885; Brighton & Hove Electric Light Co 1888. A new power station was opened next to Reed's Foundry in 1885; Brighton Corporation built a station of their own nearby and commenced operations in 1891. At first relations with the Brighton &

Hove Co were bad, and the latter took the ground from under the Corporation by threatening to cut off all its own consumers unless they signed restrictive and binding agreements to take private power and no other! These difficulties were resolved by municipalization.

3 It was built on six thousand concrete piles sunk down to 50 feet: one needed ten thousand 3-ton blows from a ram before it settled. The facing bricks were of high quality, from the now defunct High Brooms brickworks at Tunbridge Wells; Brighton 'B' contrasts with its austere and functional neighbour, which dated from the Carville steel-shed era.

4 Brighton was an obvious choice for electrical pioneering, it had a compact and wealthy society of the kind which suited early low-voltage DC systems, as in the case of London. Brighton also saw Great Britain's first electric railway (Volk's Railway 1883), and in 1933 it was the terminus of the country's first main line electrification from London. A pioneer trolleybus worked at Hove (1914), and the peculiar electric 'daddy long-legs' (rather like a small pier on four legs, running along rails on Brighton sands, even at high tide) designed by Magnus Volk, showed off further potentialities of electricity. Brighton also had electric trams (1901–39) and trolleybuses (1914 and 1939–61), and there were some early film studios at Hove and 'Bungalow Town' now part of Shoreham. Much of this enterprise was due to Magnus Volk (1851–1940) a native of Brighton and later Borough Electrical Engineer. His electric lighting installation in Brighton Pavilion (1883) was the largest single one at the time; Volk also pioneered telephones in Brighton (1882) and the town was one of the few to have a municipal telephone system. Brighton has as much claim as anywhere to be the oldest home of electricity. Volk's original railway still runs, embodying the Brighton electric tradition.

5 To the detriment of the NE later, when 50 Hz became the national standard, enforced under the 1926 Act: one of the penalties of having been a pioneer, and the 'oldest firm in the business'—of real all-purpose power generation, at least.

7, *The Battle of the Systems*

AT the turn of the century, and slightly before, an academic and industrial battle was fought about the question: Alternating Current, or Direct Current? Periods of rapid change, like the Electric Revolution, tend to generate excessively rigorous or exclusive claims, and an unwillingness to compromise among subgroups in the movement. A prior cause of this 'theological disputation' was the localist distribution pattern of power generation and distribution, which allowed one set of engineers to install AC systems, while another chose DC, sometimes in the next parish. In this way the 1888 system was intensified in its weakest part, and it became all the more urgent to stop the rot before it went too far: hence the 'battle of the systems' as it was called. By 1900 it had become clear that large-scale supply (such as the power companies were offering) was a real possibility. The economic and potent Parsons turbine had partly unbound the electric Prometheus, but the mass of heterogeneous original systems, statutorily protected, remained to trammel him still.

Leading scientists and engineers took sides in the systems debate, each tending to claim exclusive truth. Their behaviour was fully in accordance with the stimulating confidence of the new age, although it was in many ways an unfortunate complement to the distribution pattern when it was grafted on to it. The dispute was not peculiar to Great Britain. In the USA, the pioneer Edison was a firm believer in the efficacy of DC, and so too was the giant General Electric Company which sought to gain contracts for early traction and lighting systems. Indeed, GEC had been formed for this purpose by uniting the Edison GEC and its rival, the Thomson-Houston Co, in 1892. On the other hand, George Westinghouse (using Tesla's AC patents) was a firm 'AC man', and used the AC case to belabour the GEC 'electric trust' as its enemies called it. The antagonists fought in all imaginable spheres. The AC interest was furious when General Electric won the contract to install the electric chair at New York's State penitentiary at Sing Sing and equipped it for AC, thereby besmudging the AC image. The press claimed, ridiculously, that AC was 'more lethal' than DC. The chair was erected in 1889, and after a legal battle as to its constitutional correctness, despatched its first victim in 1890: 'effected with difficulty' as Haydn's *Dictionary of Dates* put it rather gruesomely. An early victory for AC was the Niagara hydro-electric system. Shortly after 1900, the battle changed its field for the prize of railway electrification, about which optimistic and exaggerated views were being aired.[1] This was the main reason for Westinghouse's move to Great Britain 1899–1902, when the subsidiary 'British Westinghouse' was set up, with a magnificent works at Trafford Park, near Manchester. British Westinghouse was ready to counter any similar move from American GEC (whose British subsidiary was called British Thomson-Houston), or British GEC, or any other taker.

In the USA matters got out of hand soon after GEC won a contract for electrifying part of the New York Central's[2] railway system around New York: on the DC system, of course. George Westinghouse won a contract from the New Haven RR,[3] which was the New York Central's own neighbour, for an AC system. The two systems differed in other ways: the New York Central was to use third-rail conductors, the New Haven overhead wires ('catenary'). Westinghouse wrote an open letter to the President of the New York Central (October 1905) while the General Electric contract was being carried out: 'Stated briefly your situation seems this: your power house and its equipment now under construction is suitable . . . for the generation of single-phase equipment . . . you know the diametrically opposed views and interests of Westinghouse and General (GEC) companies, and of the strife between them . . . there seems to be no good reason why your company should be the victim of such commercial strife'. To the *Railroad Gazette* he wrote: 'there are arranged today against (single-phase AC) many of the same men and same interests, actuated by the same commercial spirit and using the same methods and tactics as were employed by them when they began their "peculiar" opposition to AC'. He went on to criticize Frank Sprague,[4] one of General Electric's leading traction experts, who riposted: 'the charge that GEC and myself are engaged in a conspiracy to fasten upon the railway world a DC system . . . is an amusing product of imagination, or the result of hysteria . . .' Westinghouse contended that General Electric had lost the fight to become a lighting trust, and was now making a bid for traction; some replies to Westinghouse noted that he might not be without the 'commercial spirit' himself, in spite of his pretended altruism.

In Great Britain the muddle of the systems was crystallized by 1900, and both sides took up positions, hoping to prevent the spread of the other. The fight never became so intense and personal as in the USA, for example, Colonel Crompton (DC) thought that when he met Ferranti (AC) at the 'Dynamicables' dinner, that they were both too occupied with the event to do aught but 'chaff and banter'. They remained on good terms, and Crompton was an honoured and saddened guest at Ferranti's funeral (1930) when AC was carrying all before it.

The main antagonists were Crompton, Kelvin and Hopkinson (all DC); Ferranti, Mordey and Sylvanus Thompson (AC). Kelvin was an important element in the DC cause, being the leading physicist of the day; in 1907 he stated (as DC was wallowing badly): 'I have never swerved from the opinion that the right system for long-distance transmission of power by electricity is by the continuous current system.' In 1894 it was thought no small matter that Kelvin consented to open the Coatbridge[5] AC generating station of the Scottish House-to-House Electricity Co, as in 1901 he opened Neptune Bank, Wallsend, the pioneer three-phase AC station. Both stations furthered the cause of the 'other side'. However, the battle of the systems did not usually degenerate into tendentious or personal attacks, and Kelvin was interested in all forms and applications of electric power as such.

Having set the scene of the battle, its technical nature can be dealt with briefly, for this has some bearing on its outcome in later years. In generation AC enjoys the advantage that alternators (AC generators) can generate at a high voltage, whereas it is harder to design dynamos (DC generators) to do this. In transmission, AC can easily be stepped up to and down from high voltages by means of transformers. High voltage transmission saves on the size of the mains; high voltage DC could only be changed then by 'DC transformers'—really motor-generators that received high voltage DC and turned it into low-voltage, for domestic use, mechanically. At the consumer's end DC was superior for arc lights, but these began to go after 1920, and it had no advantage over AC (above 40 Hz) for lighting purposes. In the 'battle of the systems' era DC was preferred for traction and motors because of the simple, flexible operation of the DC motor. AC motors were more difficult to construct and maintain, apart from the rugged three-phase motor which ran at fixed speeds and so presented difficulties for traction.[6] The supplier liked the ability of DC to be stored in batteries against peak-hour demands, but this was only really useful so long as electricity was largely used for lighting. The huge demands of industrial and traction businesses for power would soon have flattened the modest storage batteries employed by local lighting companies. AC, which cannot be stored, needs enough capacity in generating equipment to cope with sudden demands, and this means that generators are not always working at optimum capacity speeds, hence the obsession with 'load building' in the Electric Revolution.

In some way the qualities look evenly divided, and, given the parochial nature of supply in 1900 there was some truth in this. However, once big-scale generation and industrial users became common, DC fell by the wayside and survived only in a few local lighting companies. This pattern began to emerge with the Parsons turbine/power company era, and when the great urban suppliers were called upon to deliver power on the grand scale. Long distance linkages between systems, and the coming of the 'grid' put the *coup de grâce* on DC. Before the Parsons age few people thought in these terms. A leading electrical academic, Professor (later Sir) Alexander Kennedy thought (1894): 'Possibly 4,000 hp in the station . . . had got to a point beyond which . . . they were not going to reduce anything in the cost per unit.' Ferranti believed in the efficacy of high voltage AC but his visionary power station at Deptford had a very unhappy start, and retarded the AC cause somewhat.

The Deptford station ran into considerable trouble. The 1888 Act cut down its supply area. The original London Electric Supply Corporation station at the Grosvenor Gallery was converted into an AC transformer station, but burnt itself out in half-an-hour (15 November 1890) when a switch arced by being withdrawn under load; the original mains cables were inadequate and a bad fog in 1891 (see p 164) overloaded Deptford and put it out of commission for four days. Being AC, and being unconnected to any grid, the failure of the Deptford machinery cut off all London Electric Supply Corporation customers and made most of them furious; DC could, of course, have fallen back on batteries. The London Electric Supply Corporation ran into debt and seemed to generate nothing but badwill. It paid no dividend until 1905, and Ferranti left under something of a cloud.[7]

The first DC systems generated at a low voltage (about 100V) and distributed to consumers direct, by means of underground mains, although a few used a cat's cradle of overhead wires at first, by way of fulfilling the terms of their Provisional Order in time; it was usually obligatory to bury them. These mains, because they carried a low voltage, were usually rather large and clumsy, and therefore expensive—a matter which encouraged engineers to look around for systems that could employ lighter mains. Typical low-voltage DC systems were: the first Metropolitan Electric Supply Co (1888, 110V DC, later converted to AC); the Kensington & Knightsbridge Electric Lighting Co (1887, 100V) and the publicly-owned St Pancras Vestry Undertaking. One of the first DC variations was the high voltage DC system, as tried at Colchester (1884), Oxford (1891) and Wolverhampton (1895)—in these systems DC was sent at high voltage, up to 2kV, to batteries or DC transformers which thence supplied customers' homes with low voltage current. The long distance mains were thus much smaller in cross section. Colchester placed five great batteries about its network, Oxford did it by reducing 1kV to 100V via motor-generators (it was sometimes called the 'Oxford System'), and the Cadogan Electric Lighting Co (1887) actually put receiving batteries in consumers' houses. The battery could discharge at 'house voltage' which might be anything from 12V to 96V depending on the size of the house, and the number and type of its lights. A common variant was the 3- and 5-wire DC system, by which dynamos generated at (usually) 460V and supplied various domestic circuits at half of that voltage. This required careful balancing and it was restricted to compact urban areas. Many undertakings offered both AC and DC, for example Manchester (1925) with DC (3- and 5-wire); AC and five voltages.

The first AC systems were contemporary with the first DC ones, for example Eastbourne and Brighton, the latter abandoning DC in 1887.[8] London's most famous early AC system was Ferranti's, centred on the Grosvenor Gallery, New Bond Street (1883). It was a very modest local system (2kV transformed down in houses) and was taken over by the London Electric Supply Corporation in 1887; in this way the larger firm acquired the services of the Grosvenor Gallery consultant, Ferranti, with momentous consequences for them and AC supply in general. The plant was originally conceived to light the actual gallery, but it took on other customers and eventually overhead conductors radiated in all directions from the Gallery station, swinging over the housetops from Regent's Park to the Thames. Other London AC systems were: the Metropolitan ES Co (1kV, 100 Hz); the Brompton and Kensington ES Co (2kV, 83 Hz) and Islington Vestry (2kV, 50 Hz).

AC's internal variations lay in frequency, phase and voltage. As with DC, these curious variations did not matter so long as electric power remained a local affair, but it was to cost dear in later years: frequency standardization cost £17,500,000 in the 1930s. AC won the 'battle' because it could be easily transformed to high voltages for long-range transmission, and stepped down with equal simplicity for local and domestic distribution. The transformers were non-mechanical and could be left unattended for months on end. At the common (and later standard) 50 Hz it was excellent for lighting by incandescent lamps, and it could be converted to DC for traction.[9] Later design and production of AC motors overcame the initial prejudice to them as being 'clumsy and inflexible', and industry employs them very commonly, as do many households in vacuum cleaners, clocks, washing machines, etc.

But the war was never fully over. DC still has many uses, particularly in heavy industrial work. The extensive network of Southern Region and London Transport railways employs it entirely for traction. In later years the battlefield moved to electric railways in general (see below), and the deployment of systems at various stages in the war is still embodied in the differing systems of electric traction apparent in Great Britain. Thus the outcome was victory for AC; but not conquest. In that non-mechanical generation produces DC, and in that this system may be a major power source in the future, then Kelvin may yet prove to be right.

In the same way that early and rather myopic views had caused, or overlooked, the inherent disadvantages of different railway gauges and lighting voltages, so systems of railway electrification came to display a lack of uniformity, although not so wide a one. Most electrification plans were local, and the general unwillingness of main line companies to electrify trunk routes kept it that way for many years. On the few occasions when main line electrification seemed a real possibility (1919, 1931 and 1956) government enquiries and reports attempted to establish standards binding on all railways. Owing to the continuing battle of the systems, British Railways have three systems of electrification, dating from particular phases in the battle. The original, or 'old suburban' system is low voltage DC, with a third rail conductor; the 'middle electric' dating from the inter-war period (and advocated by the 1919 and 1931 reports) is high voltage DC, from overhead conductors. Finally there is 'late electric' or perhaps 'modern' which is represented by the new standard of 1956: 25kV AC, using the industrial frequency of 50 Hz. The AC is converted to DC on the actual locomotives by means of rectifiers—an admirable solution to the old AC *versus* DC war, made possible by improvements in rectifier design. However, for all the technical sophistication of the modern AC system, there are over one thousand miles of third-rail DC in operation, mainly in the South of England. It is a well-entrenched system, and there are plans for its extension. The most likely casualty in the war is 1,500V DC, which is not a very extensive system.[10]

The Electric Revolution coincided with, and caused, largely unfulfilled and optimistic waves of electrification proposals. Before the Great War the railway electrifiers were divided (Cf. Westinghouse correspondence) between three systems: single-phase

AC; three-phase AC; DC, and the correspondence columns of the *Railway Engineer* and the *Railway Gazette* manifest this new area in which electrical dogmas were fighting it out. In a masterly and balanced summary of the 'state of play' towards the end of this controversial period (1912) a leading electrical engineer, Dr Kapp, outlined the optimum use of the various systems thus: low voltage DC for suburban work; high voltage AC for long-distance trains. He was fair to three-phase, but could not obscure the strong disadvantages of this interesting and eccentric system; in the long run his suggested solution has proved correct[11] but by accident, rather than by design.

Until the revolutionary 1956 proposal to employ industrial frequency AC on British Railways, DC had been a firm favourite with many traction engineers. It was simple, well-tried and easy to control on the locomotive or motor coach of the train. As with lighting, so with traction. Apart from semi-tramway systems, like the Giant's Causeway line, and the normal electric tramways, the first electric railway in Great Britain was the pioneer 'tube' line under London: the City and South London Railway (1890) which used 500V DC (later 550V) and third-rail conduction. London has remained faithful to this arrangement ever since, although a fourth (return) rail was added later, and the Metropolitan Railway contemplated a three-phase system at the turn of the century. This would have been an interesting, if rather bizarre variation; but how delightful to see the network of overhead conductors at Baker Street or Harrow on the Hill, or locomotives with spidery pantographs picking their way about Neasden yards! In the event, the Met had to swallow its pride and accept the third-rail system of its detested neighbour, the Metropolitan District Railway—this was by Board of Trade *fiat*, an early example of government-enforced electrical standardization, at a time when it watched helplessly as the battle of the systems raged elsewhere. The Board of Trade stipulated that both railways should use 600V DC third-rail, with a fourth rail for negative, insulated from the earth—a peculiar, although not unique, London feature which has persisted ever since. Other early urban and suburban systems were virtually all third-rail DC, including the unusual Liverpool Overhead Railway, Britain's only 'Elevated' of the American type: it thundered its way along the Merseyside waterfront for some sixty years (from 1893).

The first note of discord in this DC-dominated world came when the Midland Railway, and the London, Brighton and South Coast Railway strung up overhead catenaries and powered trains by high tension AC (1909). The Midland line was isolated, a small system in Lancashire;[12] but the Brighton Line electrified track ran into the heart of London at Victoria and London Bridge stations. The LBSCR electrical mandarin, Philip Dawson (later Sir Philip) spoke confidently about electrifying an extensive suburban network and very probably the main line to Brighton and Worthing. It seemed as if the New York Central-New Haven war might start in Great Britain, but the difficulty of raising capital for either AC or DC extensions on any scale kept the argument largely academic before 1914.

AC suffered a strong disability during the Electric Revolution when it was used for traction purposes; AC traction motors (then) worked best at low frequencies; 25 Hz was normal, although $16\frac{2}{3}$ Hz was used on continental Europe. These frequencies were too low for lighting, so the railway AC had to be especially generated, or reformed by a frequency converting motor from higher periodicities. The LBSCR bought 25 Hz power from Deptford (London Electric Supply Corporation); the MR generated its own, from a gas engine plant at Heysham Harbour.

After the Great War there was another burst of railway electrification speculation, the most promising example being in the North: the York-Newcastle electrification scheme of the North Eastern Railway, in which Merz had a hand.[13] This would have been Britain's first trunk line electrification, and would have used 1,500V DC; third-rail conduction was suggested at first, overhead later. Down South, the Brighton was talking of electrifying again, and the battle warmed up, since one Brighton neighbour (the London and South Western Railway) had begun third-rail DC operations in 1915. Mainly because of the NER plan, but also because of others that seemed 'in the pipeline'[14] the government requested that all plans be frozen while it investigated the matter of standardization. The request came from the new Ministry of Transport, anxious both to avoid a fiasco, and to exert its influence. The Kennedy Report (1921) favoured 1,500V DC, but it came too late to help the North Eastern Railway plan which was already on the rocks because of economic depression in the North East. Ten years later there was much talk of railway electrification to relieve unemployment, and the standardization question was reopened. The Weir Report on railway electrification (same man as the 1925 Report, but a different subject) appeared in 1931. It suggested spending four hundred million pounds on electrification, and holding to the Kennedy standard of 1,500V DC, overhead. Sir Philip Dawson, now the doyen of railway electrifiers, and loyal to AC as ever he was, reported in the *Electrical Review* that AC was now being adopted very widely, for example by the Pennsylvania RR in the USA, and that DC was obsolescent for main line work. Conservatism triumphed, however, and the Weir suggestion was embodied in a Statutory Instrument, thereby legally enforcing 1,500V DC. Some mileage was so electrified, in fact, notably the Manchester-Sheffield main line through Woodhead tunnel, but the post-1956 burst of activity vindicated Dawson; it was all done on the AC system, and the DC order had to be expunged.

A variation on the railway 'battle of the systems' (and also resolved by the state) was the question of generation, and whether a railway company, or a public supplier should undertake this. To Merz, greatest opponent of railway-generated power, one of the main virtues of electric traction was that it eased the load problem of large central stations, in the way that the North

Eastern Railway Tyneside lines helped Carville, his early *chef d'oeuvre*. Dawson similarly arranged for the LBSCR to take current from Deptford. Generally, however, this advice was ignored, and railways set about duplicating the already patchy system of generation by building their own power stations, some on the grand scale: the Metropolitan Railway at Neasden, the London and North Western Railway at Stonebridge Park; the London and South Western Railway at Durnsford Road, Wimbledon; and the Underground Group of London at Lots Road, Chelsea. Merz made some sharp criticisms of this practice from his official vantage point as a sub-committee chairman on the Haldane Committee (see p 44) in the Great War, drawing attention to the unnecessary proliferation of power stations about London, and the generally poor siting, away from adequate water supplies, of these 'rail-bound' establishments.

Already the state had encouraged railways to buy bulk power from established supplier, for example by the 'Kitson Clause' in power company acts. Now it stopped any further duplication with the 1919 Act. The Electricity Commissioners set up by that Act had to give their consent to new power station schemes, and in the case of railways they simply refused. The first to be thus stopped was the proposed South Eastern and Chatham Railways' station at Angerstein's Wharf, on the Thames. One by one traction power houses were shut when expensive renewals did not seem justified. Once again, rationalization, personified in Merz, won in the long run, and the battle had been fought out within the Electric Revolution.

A table will most conveniently display the variety of systems operating in Great Britain during the Electric Revolution, and some related matters such as the heavy dependence on foreign engineering and technology which prevailed in Edwardian times. Indeed, much early electric rolling stock was clearly American influenced (and occasionally constructed),[15] with broad clerestory roofs, generously proportioned driver's windows and genuine

PRINCIPAL RAILWAY ELECTRIFICATIONS IN GREAT BRITAIN, 1883–1915

Railway	*Year*	*Type at opening*	*Foreign influence, if any*
Volk's Railway, Brighton	1883	50V DC 3R	
Giant's Causeway Tramway	1883	250V DC 3R	Water turbines by Allcotts of NY, USA, motors by Siemens
City & South London	1890	500V DC 3R	
Liverpool Overhead	1893	500V DC 3R	Later (1900) equipment by British Schuckert & Westinghouse, USA
Central London	1900	550V DC 3R	Locos by American GEC; BTH motors
Great Northern and City	1904	600V DC 3 and 4R	BTH equipment
Yerkes Group[1]	1905–7	550–600V DC 3R/R	Much stock by US, French and Hungarian makers/BTH motors
Mersey	1903	600V DC 3R/R	Equipment by US and British Westinghouse
Metropolitan	1905	600V DC 3R/R	British Westinghouse power stn, BTH stock
Lancashire & Yorkshire	1904–13	630V DC 3R[3]	
North Eastern	1904	600V DC 3R	BTH contract for electrification
London Brighton & South Coast	1909	6,600V AC o/head	AEG contract, Winter-Eichberg motors BTH and Siemens
Midland	1908	6,600V AC o/head	British Westinghouse stock
London & South Western	1915		Siemens and BTH[4]
London & North Western[2]	1914–15	630V DC 3R	Siemens and Oerlikon stock
NER Shildon-Newport	1915	1,500V DC o/head	Siemens

1 Charles Tyson Yerkes was an American traction financier who was largely responsible for building the three principal London tubes, and electrifying the District. Eighteen million pounds was spent on this, much of it American. The tubes were: Great Northern, Piccadilly & Brompton (Piccadilly line) 1906; Baker Street & Waterloo (Bakerloo) 1906; Charing Cross, Euston & Hampstead (Northern) 1907. The Metropolitan District was electrified in 1905.

2 Including parts of the North London Railway, worked by the LNWR.

3 Liverpool-Southport, 1904. Some lines around Bury were electrified in 1913, variously at 3,000V DC o/head latterly 1,200V DC third rail.

4 By 1915, British Westinghouse was largely in British hands, and Siemens was in the hands of the Custodian of Alien Property.

Key

3R: third-rail conductor
3R/R: third-rail conductor with fourth rail for return
3 and 4R: [GN&C only] two conductor rails, outside running rails

o/head: overhead wire conductor
BTH: British Thomson-Houston
AEG: Allgemeine Elektrizitäts Gesellschaft

American 'car builders expanded roman' lettering on the body-work. The early Mersey or London 'District' stock would have looked quite at home on the 'South Side El'.

Although a great deal of suburban electrification took place during the Electric Revolution, only at the very end of it was a main line[16] so treated: London to Brighton 1933. The high hopes of Westinghouse and American GEC that Britain would go in for large scale main line electrifications were not to be realized for sixty years. In the meantime, a rather heterogeneous collection of railway electrics kept the battle of the systems smouldering on into modern times.

NOTES

1 Electric railways have many operating advantages over steam, or diesel ones. But they are immensely expensive to build, and require heavy and continuous traffic to justify any investment. High interest rates are, initially, their greatest enemy—one reason why not building them in the 1930s when rates were around 2½ per cent, and unemployment high, was very unfortunate.

2 Correctly the New York Central and Hudson River Railroad, then.

3 Correctly the New York, New Haven and Hartford Railroad.

4 Electric traction's great innovator in its early years, see *Men in Business*, ed William Miller.

5 An unusual little plant, operated initially by gas engine-driven alternators. Early alternators were 'the very devil' to get working in parallel as an engineer put it; gas powered units, with uneven turning moment made it more difficult still. Each Coatbridge engine had three fly-wheels, but to no avail—the system took up steam power after two years. Some gas plant, particularly later and larger direct-coupled sets ran quite well: eg at Accrington and Heysham Harbour.

6 There was a considerable network of three-phase in Italy, and a few scattered systems elsewhere: the first Simplon tunnel electrification, and the Great Northern Railway (USA) Cascade tunnel. Three-phase needed two overhead wires (for two phases) and a bonded running rail for the third. Junctions, of course, were fiendishly complicated. Trains kept good time, running at their fixed speeds: three-phase electrifica-

tion had as much to do with making Italian trains run on time as ever Mussolini did.

7 See Appendix II, Ferranti had left before the fog fiasco.

8 One AC advocate went over to the DC camp: J. E. Gordon, following his harrowing experiences with early alternators at Paddington station. See p 64.

9 By rotary converters in substations—later by rectifiers. It then went into a third-rail conductor or an overhead. Rectifying on locomotives was very unusual until modern times, although there was some, usually by mechanical means, adding thereby to the loco's weight.

10 Being converted in 1970 on the old Manchester, South Junction & Altrincham line, for example.

11 Glasgow's suburban 6,600V AC and the London-Bournemouth-750V DC notwithstanding.

12 Lancaster, Morecambe and Heysham line.

13 See *The Electric Railway that Never Was; York-Newcastle* 1919 by R. A. S. Hennessey.

14 Including the Great Western Railway's East and West Valleys routes in Monmouthshire. The GWR occasionally toyed with electrification schemes to bring down the price of best Welsh steam coal, which it bought in large amounts.

15 The original Bakerloo tube carriages (built by the American Car Foundry Co) bore the legend: 'Baker Street & Waterloo RR', no less.

16 At the other end of the scale there were electrifications of small sidings, colliery lines, etc, where this proved economical. Not surprisingly some power stations (like NESCO's Dunston) were so served, also the considerable network of the Harton Coal Company near South Shields. Oddly, two mental hospitals had electric lines: Hellingly in Sussex, and near Leek, Staffordshire.

8, The Local Lighting Company—'K&K' and the Frome Undertaking

AT one level the story of the Electric Revolution is concerned with millions of units, of consumers and of gigawatts; but for much of that fifty years, to many people, it was the rise and development of the local electricity company, the 'light company' or the town electricity works. The manager and his staff, the meter readers and the showroom salesmen were fairly familiar figures. Lest one becomes lost in the world-historical struggles of engineers and political titans, the example of two local undertakings will redress the balance.

Down below (and behind) the Royal Albert Hall in London are the Albert Vaults, where, fifty years ago the 'K&K'—the Kensington & Knightsbridge EL Co had a distribution department. Impeccably maintained motor generators bearing the legend *Maschinenfabrik Oerlikon* in high-relief turned the 5kV AC power from Wood Lane power station into 400V DC and 200V DC for local consumers. These included the great Kensington museums, the Knightsbridge shops from the mighty Harrods to Madame Fifinella, *haute couturier*; and the houses of such social families of high standing as Goschen, and Michie.

In the Albert Vaults station, Mr Gleave (ex-Royal Flying Corps) tested the undertaking's meters, mainly manufactured by Ferranti, and Chamberlain & Hookham; Mr Anscombe and a boy went out to do repair work for customers in distress. The museums could generally handle their own crises, but not even the houses of the grand ran to a permanent electrician on the establishment. After Mr Gleave had tested the meters, the Cubitt brothers installed them on domestic premises and tested installations. Over the whole affair presided the distribution engineer and manager, Mr. G. C. Weston, assisted by his foreman. Mr. Fortescue, a generator engineer, doubled as Shop Steward: all workers at Albert Vaults were members of the Electrical Trades Union, and in their case were part of LSE 4 (London Station Engineers) branch.

In late 1918, LSE 4 made trade union history by a dramatic stoppage which demonstrated both increasing trade union power, and social dependence on electric power. The post-Great War period was one of considerable militancy and radicalism, and LSE 4 satrapy of the ETU was called upon to strike a blow for the cause of labour. The pro-labour *Daily Herald* had arranged a socialist meeting in the Albert Hall, but the Hall council later backed down on this agreement, it was believed under government pressure. The ETU requested LSE 4 to bring pressure on the council by industrial action, and some of the Albert Vaults engineers removed the Hall's fuses. Faced with the prospect of a darkened and inert Hall *ad infinitum*, the council surrendered, and the *Herald* meeting went ahead as planned. George Lansbury, a leading Labour Party politician boasted: 'the men of the ETU who darkened the Albert Hall were quite able to shut down the supplies of current to all London'.[1] Lansbury's accurate, if slightly impolitic observation demonstrated the power not only of trade unionism, but also of the great potency of the electricity supply industry.

In the 'K&K' station at Albert Vaults everybody knew everyone else, and many of the customers as well. Even great corporations have some departmental life, of course, but the K&K had more of the *esprit* of a family firm; although the kind of localism that it represented played havoc with the plans of sophisters and economists, its passing leaves society all the poorer.

K&K was in the Crompton empire, and dated from the early days of London's electricity. Many provincial undertakings came later, and were isolated examples of local enterprise.[2] One such was the Frome Undertaking, owned by the Urban District Council, and managed for it by Edmundson's Electricity Corporation. In 1900, Frome, Somerset had a population of 12,000, and obtained a PO from the Board of Trade in order to light and power the neighbourhood by electricity. The UDC interviewed seventeen electrical engineers and appointed F. Medhurst as their consultant; a sub-committee was formed to supervise the proposed undertaking, and the politics started. The UDC chairman, H. T. Rawlings, with four other councillors, wanted Frome UDC to manage the undertaking itself. On the other hand, a big ratepayers' pressure group was very sceptical of this aspiring municipal enterprise, and whipped up hostility to the project. Far better to leave the matter to engineers and managers who know about such things, they claimed, than to bureaucracy and conciliar punctilio—which would eventually lead to a 'burden on the rates'. A compromise was arrived at, whereby Edmundson's would build the station for £20,156 5s 0d (£20,156·25p) and manage it. Frome UDC would own it, and thus satisfy municipal pride; Edmundson's would run it, and satisfy the hard-faced businessmen who scoffed at amateur management.

The UDC commissioned the building of the station, and the architect (J. Armitage, of Broad Sanctuary, Westminster) designed a most pleasing building for the town. It was finished in 1903, fashioned out of local stone with Bath stone quoins and dressings, vaguely reminiscent of the work of H. H. Richardson, rugged but tastefully neat. The dominant feature was a 110 ft chimney in red Cattybrook bricks, with white brick bands and topped with a cornice of Staffordshire blues. Within the station there were two Parsons 150kW turbo-generators, and one 75kW set; they rested on plinths, surrounded by a floor of glazed terazzo.

At one end there was a switchboard, surmounted by a clock entwined in the most intricate iron scrollwork; like the street lamps, this was by the Singer Art Metal Co, a local firm.

On 1 July 1904 private supply started to consumers, at seven pence and three pence (3p and 1p) (two-part tariff) for light; fourpence and one penny ($1\frac{1}{2}$p and $\frac{1}{2}$p) for power—customers could also choose a flat rate at sixpence ($2\frac{1}{2}$p) for light, two and one half pence (1p) for power: all prices per unit. Public street lighting started one month later. The undertaking was typical of many, in its tariff structure, in the municipal pride that it created, in its effects on the local gas undertaking, which invested in better equipment and reduced its charge from three shillings and fourpence (17p) per thousand cubic feet to three shillings (15p). The undertaking started out supplying just over three thousand 8-cp lamps, within a year its load was nearly twenty thousand of them. It powered 102 motors in various factories, from $\frac{1}{4}$-hp to a 35-hp model in Butler and Tanner's printing works.

A year after the supply started, an anniversary dinner was held at the Bull Hotel, Frome—a dinner full of significant incidents and symbols that go to make the Frome undertaking the perfect microcosm of the British power problem. The dinner was dubbed by the press the 'yellow dinner': Mrs Herbert Ames, wife of the clerk to the UDC, had arranged fifty yellow-shaded electric lights, and clusters of yellow carnations. At the table junction a fountain played on some lilies, and electric fans worked around the room: all very *art nouveau* in conception. The speeches revealed some deep-seated and incompatible political confrontations. E. R. Trotman admitted the undertaking did its job well, but the street lamps were more in number and more costly consequently than gas lights. H. T. Rawlings displayed similar Bourbon tenacity; after congratulating the good works of Messrs Edmundsons he reminded all present that Frome UDC had a reversionary right of purchase after seven, ten and fourteen years, and clearly he still hankered after UDC management. The leading local businessman, Lanfear R. Tanner quickly replied that 'electricity requires delicacy of touch and it is unwise for any but experts to undertake it [supply] if it is to have success—if they [UDC] undertook it themselves, the Lord help them'. The correspondent of the *Electrical Review* was present and tried to soothe ruffled feelings by saying that if the UDC did undertake the matter themselves, and that if they kept on Edmundson's agent and manager (Mr Michell Nicholl) then the Lord *would* help them. One more row remained, of that sort of provincial cyclone found in Gogol, very real, however, to its participants. The editor of the *Somerset & Wilts Journal* was (as doyen) asked to reply to the toast 'The Press', which he did. This was too much for the *Somerset Standard* which let go in its next number: 'We who have dared to offer any criticism on the scheme were completely extinguished by the blaze of the electrical enthusiasts . . . and the mutual congratulations; . . . we were positively ignored while the health of our contemporary and a London periodical were drunk in bumpers.'

In the Frome affair one sees all the Electric Revolution controversies in miniature, except perhaps the battle of the systems. Some wanted municipal enterprise; others wanted it to be private. Local patriotism fought, often within the same breasts, with fiscal caution, vested interest and political rivalry. How many, one wonders, realized at the yellow dinner, that the entire controversy would be academic within fifty years, and that the electricity supply pattern of the whole kingdom would become national, immense and integrated? Events have moved so quickly that the local lighting undertaking seems part of another world, like the iron scrollwork at Frome and the Oerlikon motor-generators in the Albert Vaults.

NOTES

1 Perhaps this kind of talk led to the amendment in the Electricity Supply Act of 1919 (thus modifying trade union law under the Conspiracy and Protection of Property Act, 1875) which made power strikes illegal. This statutory exception to the general freedom enjoyed by trade unions in taking industrial action was much quoted during the 1970 'go-slow' that demonstrated the dependence of modern society on electric power.

2 Some were very local indeed, for example the man from Carperby, Yorkshire, in Graham Turner's *The North Country* who said : 'In 1925 the only electricity was provided by a joiner with the help of an Austin car engine—it made a terrible noise.' However, such power plants and the very small statutory ones did much to sustain a feeling of community independence, impossible with integrated grid-systems : another example of 'liberty or efficiency', perhaps?

9, *Hydro-Electricity—Politics and Power III*

The Systems

HYDRO-ELECTRICITY is generally associated with the cyclopean dams of the USSR or North American: Dneproges, Ontario Hydro and the Hoover Dam. To the extent that it appears in Great Britain it is usually regarded as modest in extent, and very much 'Celtic Fringe' in location, also it is supposed to be modern, and largely post-1945. While it is certainly not extensive (it provides but 2 per cent of Britain's electric power), it does have a very respectable antiquity—the 1881 Godalming scheme was hydro-electric—and it is to be found outside the Highlands of Scotland and Wales, although rather more on the wetter, Western side of the Penine chain. The Electric Revolution saw the rapid adoption of hydro-electric schemes, which were sometimes accepted as a means of freeing local (often very local) power needs from the 'thraldom of coal'. Thus, although early hydro-electric power was not an energy producer on a large scale, it does exemplify the eagerness of consumers to avail themselves of the new energy as quickly as possible, and to not a few remote houses, factories and towns this was a real leap forward in standards of convenience. The big schemes came later: the Falls of Clyde; Galloway and the North of Scotland.

Private houses, especially of the rich, were frequent users of small hydro-electric plants. The earliest such installation (and a candidate for 'first hydro-electric plant in the world') was 'Cragside', near Rothbury in Northumberland, the remote Victorian-Gothic eyrie of Sir William Armstrong, the engineer and armaments manufacturer. He was a personal friend of Swan, who installed a lighting system at Cragside in 1880. It was to inspire many others. At the other end of England, Rudyard Kipling built his own hydro-electric plant at his Sussex home, Batemans. The contrast of the setting was complete: Batemans was placed in soft green countryside, set off by some exotics planted by Kipling himself. On the other hand, Kipling had a genuine delight in things mechanical, like Armstrong, and placed a water turbine in a converted water mill, which ran from 1902–27 and could light twenty 100W bulbs at 100V. Kipling had an affection for the system, and wrote: 'He [the turbine] worked for a quarter of a century' and when it was gone he 'never got anything so faithful'.[1] Installations of this type were more common and popular among the 'big hoose' type of residence in Scotland, where water power was plentiful and regular. Such a scheme could, for example, power a house near Loch Awe (1927) and render it pretty nearly all-electric: an immersion heater in a 60-gallon water tank, a drying room, electric irons, vacuum cleaner, etc, all in the (then) far away Highlands. Hydro power similarly lit Castle Forbes (of Lord Forbes); and Shielbridge House of Kenneth M. Clark, chairman of the Aluminium Corporation, itself a heavy user of hydro power, and which used its expertise to install the chairman's own system in the mid-1920s.

In a few instances the local laird or squire would extend a transmission line to serve a nearby town or village. This was one of the last examples of squire-paternalism, and demonstrates how quickly the Electric Revolution was moving. In this way, the Marquess of Graham extended his private system to serve Brodick, Arran; and Colonel Vaughan Wynn lit the North Wales town of Corwen (1923).[2] A related type of scheme, in Scotland (1902), arose when Major E. W. Blunt, RA, husband of the Countess of Cromartie, electrified Dingwall and Strathpeffer Spa. He had a great reservoir dug out of the local peat, into which flowed the Allt Gleann Sgathaich; the *Electrical Magazine* correspondent was excited by the power house 'situated in a wild region close to the frowning Raven's Rock', Similarly, in 1927, Sir J. Shelley Rolls, Bt, initiated a small hydro-electric plant at Avington, near Winchester, which served his own house and the nearby villages of Easton and Itchen Abbas—all from a plant using a mere three-foot head of water! These local gentry-inspired systems were last in the long run of local welfare, which had started in feudal times; such enterprises seem to belong to the remoter past in the age of the welfare state, but they serve to remind us of how enterprising and conscientious many local landowners were.

The original industrial application of hydro-electric power was the Giant's Causeway Tramway[3] in Ireland. This line picked its way along the coast and headlands between Portrush and the weird basalt formations after which it took its name, fully in keeping with the rugged and romantic surroundings so often associated with hydro-electric power in the British Isles. The Giant's Causeway Tramway was as great an innovation for its time as any in history, a trail-blazing pioneer, 'the first innovation of its kind yet used', as the *Telegraphic Journal* observed.[4] Among the shareholders was Sir William Thompson (later Lord Kelvin, the scientist) and Dr William Siemens, whose firm took charge of the hydro-electric machinery installation (two 45-hp turbines, at first). On opening day (28 September 1883) Thompson japed about, holding an unearthed dynamo terminal in his left hand, while proffering the right one for a shake, thereby imparting a 250V shock—what could better illustrate the primitive ebullience of the electricity industry and its protagonists in those dawn days?

The Giant's Causeway Tramway, however, remained isolated and untypical,[5] as did the occasional factory or mill powered by hydro-electricity. A more pregnant development took place at Cwm Dyli in North Wales (1903–4) where a fair sized hydro-electric station was erected to power nearby slate quarries[6] with three-phase 50 Hz current, by tapping lakes Llydaw and Glaslyn which lay, conveniently, in volcanic bowls at about 1,400 feet above sea level. This happened a long time before environmental planning became necessary, for Great Britain had not yet become motorized, congested and overcrowded. The Cwm Dyli station beamed out cheerfully from below beetling crags, built in the best Norman-Shavian suburban Edwardian, the 'public style'. It could have been a railway station, or even a generously proportioned convenience. How delightful was an age in which such development might be easily forgivable! Now we are serious and long-headed about siting and design, and doubts have come crowding in.

Cwm Dyli was built by the North Wales Power and Traction Co in order to serve some nearby quarries and a proposed aluminium works—also a narrow gauge three-phase AC railway which was never built. In 1907 the Aluminium Corporation built the Dolgarrog hydro-electric plant, also in North Wales, in order to power its aluminium works as cheaply as possible, and so to compete with water-powered continental competitors. An earlier British attempt to refine aluminium by thermally-generated electricity (Patricroft, near Manchester, 1890–4) had been ruined by continental competition; it could only offer the metal at £1 per pound, as opposed to four shillings (twenty pence) per pound. The Aluminium Corporation scheme settled down after a period of adjustment, and it is interesting to note that it was persuaded to open a factory far away in Wallsend, where NESCO's aggressive load-building policies supplied it with ample and cheap thermally-generated power. The North Wales Power and Traction Co worked closely with the Aluminium Corporation (which bought a majority share in the electricity company in 1918), and bought surplus power from Dolgarrog for its own network. In later years this grew to a large size—Colwyn Bay UDC's undertaking was connected in 1923, and other North Wales towns bought the company's power either as a substitute, or as a booster, for their own generating stations: Llandudno, Penmaenmawr and Hawarden. Eventually Wrexham and even Crewe, over the border in Cheshire, were connected to this enterprising water-powered system. The most ambitious North Wales Power and Traction Co enterprise was to build the Maentwrog hydro-electric station (1928) which had twenty-four thousand kW capacity on opening.

The Dolgarrog plant suffered a tragic disaster in 1925, Great Britain's only major catastrophe connected with the electricity industry. Part of the ground near the base of the dam at Llyn Eigiau broke away, and three million tons of water burst out, rushing down to the valley of the River Conway. It killed sixteen people, including women and children; flooded the aluminium works and left a dismal, sodden trail of mud and rocks. The Inquest revealed a sorry tale of poor workmanship and contracting some eighteen years before. Unfortunately, the work had been suspended during the early difficulties of the aluminium scheme. A new contractor took up the work for a re-organized Aluminium Corporation, so nobody could say for certain who was responsible. However, witnesses were certain that the ground had got very soggy before the accident, suggesting heavy seepage under the dam; a tendency likely to occur with insufficiently underpinned dams. At about 9.30 pm on 2 November 1925, the water burst out from under the dam and flooded into the smaller Coedty reservoir below Llyn Eigiau; Coedty's dam burst under the onrush, and the combined water-avalanche crashed down to the Conway valley.

In Scotland, hydro-electric power was similarly developed by the aluminium industry. Foyers, Inverness-shire (1896) was designed to make two hundred thousand tons of aluminium per year; it was built by the British Aluminium Co which also set up plants at Kinlochleven (1909) and Lochaber (1929–34). These were built under private agreements with local landowners, and although there was some high feeling about these early and modest assaults on the environment, there was none of the major public outcry and fury that characterized much Scottish hydro-electric development between the wars.

In addition, there were some English and Irish hydro-electric schemes of note: Worcester (1894); Keswick (1896); Chester (1913); Tisbury (1922); York (1923); Ringwood (1925); Bandon, County Cork (1925) and West Devon (1937)—the massive Shannon scheme (see Chapter 11) was constructed 1925–9. The Chester scheme, on the River Dee, was a masterful exercise in how to eke out the maximum energy from unpromising hydro-electric material. The plant was served by the low annual average fall of six feet, and the turbines rotated at fifty-five rpm; however, it was cheap to run, averaging £2,250 per year (1913–20), and its average output was 220kW. Its designer, S. E. Britton (Chester's city electrical engineer, and sometime pupil of Colonel Crompton) later suggested a fuller development of the Dee, for forty-seven miles up to Llangollen, in which a two hundred and twenty-foot fall would power sixteen stations and generate sixty million kWh per year. Nothing came of this plan, although the York Corporation hydro-electric works at Linton Lock on the River Ouse were based on the Chester method.

The Tisbury enterprise was half financed by local residents, and half by the combined efforts of Edmundson's Electricity Corporation, and its associated Salisbury Electric Light and Supply Co. The resulting Tisbury Electric Supply Co was therefore part of that financial network emanating from the 'windy city', mentioned on p 41. When it started, the Tisbury undertaking lit 'thirty homes, four places of worship, the police station and workhouse' with 210V DC—a modest undertaking, like most English hydro-electric schemes. However, the confines of Dartmoor displayed English hydro-electric enterprise in a more promising way, although still a modest one in a wide frame of reference. Devonian hydro-electricity has a venerable history, growing as it does, out of the time-honoured water wheels of the Western mining industry. In the 1880s a small plant was set up to light Okehampton, shortly followed by another at Chagford (using a modified water wheel of the classic type). Both of these were later taken over by the West Devon Electricity Supply Co, which constructed a reservoir and generating plant at Mary Tavy, the whole being England's largest scheme (1937).

The Ringwood installation (Ringwood Electric Supply Co) worked on a four-foot fall and drove two fifty horsepower turbines; it cost twenty thousand pounds and was honoured with an official opening by the Minister of Transport, the Rt Hon W. W. Ashley, the local MP and responsible minister for electricity supply. In his speech he chaffed in a pleasant way, hoping that the trout fishing would proceed as normal now that 'Father Avon is being called on to a job of work'. He also used the occasion to declaim in an expansive way about the national electricity problem. It will be remembered that this was the time when Baldwin's government was preparing the way for reform in electricity supply: the Weir Report was four months off, and the 1926 Act just over a year ahead. The public was being prepared, in so far as it needed to be, by ministerial and other speeches. He liked the idea of a station like Ringwood's, which operated continually, tailoring supply to demand, whereas 68 per cent of plant often lay idle in Great Britain's 541 thermal stations, and as a result of our expensive electricity, the British consumption of electricity was only two hundred units *per capita* per year, compared with five hundred in the USA, and seven hundred in Switzerland. He left the audience, and the assembled reporters, in no doubt as to the attitude of HM Government as to the question of how to organize electricity supply.

The Snell Committee on Water Power Resources (for the Board of Trade, 1918–21) thoroughly investigated the hydroelectric potential of the entire kingdom, sanguinely detecting and reporting on resources in unlikely places: the Wilts Avon, for example (catchment area 674 square miles) generated six hundred and fifty kW, mostly in tiny plants. In 1920, the West Riding of

Yorkshire had six hundred water power sites, developing seventeen thousand horsepower, but the individual plant was typically very small. In general, England's topography and drainage patterns do not favour low-cost hydro-electricity, whereas North Wales had high rainfall, high places and impervious rocks—a more ideal locale for water power.

As a source of power, hydro-electricity came early in the day, but was only fully exploited much later. After 1920 it enjoyed a renaissance in Scotland, due in large measure to Scottish enterprise and engineering talent, even if some of the requisite capital came from the other side of the border. The three main developments were: Falls of Clyde (1924–7) see figure 16; Galloway (1931–6) and the North of Scotland (1943 to date). There were also some smaller ones: Bught; (Royal Burgh of Inverness 1929); Loch Luichart (Ross-shire Electric Power Co, 1929); Rannoch 1930 and Tummel Bridge (Grampian Electricity Supply Co 1933).

The Galloway scheme, of the Galloway Water Power Co, was very much the brainchild of William McLellan, Charles Merz's partner; he was one of the leading exponents of hydro-electric power between the wars. He had already made an extensive hydro-electric survey of North Wales for Lazards, the Merchant Bankers who were contemplating backing some schemes. He was consultant for Maentwrog. On the other hand he was a man of Galloway, and knew that raw, wet and remote province intimately. Here, he decided, was the main hydro-electric potential for Southern Scotland (1923), but it took many years to mobilize the requisite £3 millions[7] and the necessary political support to overcome vested interests and others. These were the years of the great depression, and Scotland was very hard hit. The Galloway scheme was 'an act of faith, almost the only long-term Scottish development to emerge from those bleak times'.[8] McLellan put it another way, with dry humour: 'any disaster would wash the bones of my ancestors into the sea'. It was the first British system of hydro-electric power designed to work in conjunction with orthodox thermal stations, a turn of events made possible by the erection of the national grid at the same time. McLellan did not live to see his creation finished, for he died in 1934 at the age of 60, worn out by his endeavours. It was also, in a symbolic way, the swansong of early electricity: no more private hydro-electric schemes were to be made, the future belonged to the public corporation in matters as expensive and controversial as this. It was truly an 'Act of Faith' comparable to Ferranti's Deptford, or Crompton's offering electricity to the Vaudeville Theatre in seven hours flat.

Few historical 'ages' are perfectly discrete, and if the central feature of the Electric Revolution—faith and enthusiasm—lived on anywhere after 1930, it was in Scotland. The North of Scotland scheme develops still, and has grown since it was first guided by Sir Edward MacColl.[9] The political origins of the scheme are mentioned later (see p 102) but it is germane to note here that it was seen as social as well as economic when it was started (1943) in an attempt to arrest the population decline of Britain's far North. It has built four hundred miles of roads, as well as developing fishing, attracting industry and above all, putting up fifty-six main dams. It is sometimes said that in the age of centralized megalopolitan culture, fresh and positive life only throbs 'out in the sticks'—with modified Londonism for the provinces. Perhaps; but it can be asserted quite safely that the spirit of Crompton, Kelvin and Ferranti survives in Caithness and Wester Ross, embodying the once axiomatic assumption that 'electricity is progress'.

Politics

The developments described above all took place in an age of rapid transition which they sped on. They demonstrate the political and social problems which later became very familiar, and fashioned thereby certain attitudes and responses which we have come to take for granted. The shifting shoals of politics, the sharpening fear of industrial despoliation, the 'regional problem' are among the more important side effects of hydro-electric development.

Very few hydro-electric schemes were ever free of political trouble of some sort; indeed it became ever more vicious and intense until the 1950s, by which time hydro-electric power had become *fait accompli*, and demonstrably not the menace that many had feared, or pretended to. Foyers (1896) and Kinlochleven (1909) mark the break between innocence and furore, and as such the quickening pace at which the Electric Revolution had moved in thirteen years. In the technical sphere, Foyers manifested certain trends that were to become an economic and political problem. The Foyers generating plant was Swiss (by Girard Oerlikon), precision built and superbly engineered, an example of the way during the 1880s that Britain had passed a 'great divide' (as Archdeacon Cunningham described it) and was being overtaken in certain traditional fields by rivals. When Kinlochleven decided to employ steel pipes (Foyers had used iron ones) welders had to be got from Germany: no British welder could do the job. On the other hand, the heavy work at Kinlochleven was performed by Irish navvies (four pence three farthings—2p—per hour), one of their last appearances *en masse*, ending a great tradition which had started in the canal and railway ages. The Kinlochleven camps were brutal and ferocious,[10] but although this gave rise to adverse local comment, the main and most bitter objections had a contemporary ring: it was objected that the smelter fumes would denude the nearby countryside, and that the whole scheme was an ugly intrusion. There had been loud and eloquent objections to railway-building seventy years before this, when Wordsworth and Ruskin had fought earthworks, tunnels, viaducts, etc, that threatened to carve up the Cumbrian Fells and Derbyshire Dales. But the railways had weathered and matured, and become unobtrusive features of the landscape. Kinlochleven marked the start of a new era: of resistance to expanding industrialism manifest in factories, pylons and so on, in hitherto very non-industrial parts. The Snell Committee on Water Power Resources 1918–21, suggested a number of places with hydro-electric potential, but efforts to develop them ran into Kinlochleven-type opposition, of a far more organized and virulent type, and not all of it emanating from altruistic nature and animal lovers, either.

The Falls of Clyde scheme managed to get through Parliament by agreeing to the most careful design and siting of its stations. The backers of the scheme[11] were indeed fortunate, for later ones had a tougher passage. A scheme for West Scotland was thrown out in 1929, and reintroduced as the 'Caledonian Scheme' it was rejected again in 1936, 1937 and 1939. The Glen Affric scheme was killed in 1929 and 1941; McLellan had to struggle for five years to get his Galloway plan moving. The political composition of hydro-electricity's antagonists is illuminating. Supporting the Caledonian scheme were many Scottish MPs who saw in it a chance to arrest depopulation; against it were English MPs who spoke of amenity, and Welsh MPs who wanted the carbide industry (which the scheme would power) for themselves. Here, in microcosm, lay some basic problems of the later twentieth century. The affluent like to visit, or enjoy the possibility of visiting, quiet rural areas: preferably empty ones. The natives do not live by rugged scenery alone, and want a chance to be affluent. Consequently, an alliance springs up between Mammon and the crofters. Nor was the Welsh attitude so deplorable as some believed. Wales was as acutely depressed as Scotland, any and every crumb of comfort had to be fought for very hard.[12] McLellan's Galloway scheme had to contend with the powerful Mining Association (the coal owners' lobby) which claimed that the already weakened coal industry would suffer from his plans. The conservationists massed again, but were bought off by the insertion in the Galloway Water Power Act of a clause: 'regard shall be had to the preservation of the beauty of the scenery'. Two other interests weighed in, and were defeated. There was the 'fear of Ayr being robbed of its birthright'—Lt-Colonel Moore MPs lurid reference to the possibility of droughts in Ayr resulting from McLellan's plans. Parliament was assured that this would not be the case. The Labour Party opposed the bill on doctrinaire grounds: the Galloway scheme was private, and therefore 'exploitation'; electricity should be kept 'in the hands of the people'. A few months after McLellan's success against his opponents, the Grampian bill failed in the House of Lords because of an alliance between the Mining Association, and the owners of local fishing rights. Although such interests could not stem the tide for ever, they did succeed then and so they display to later generations how much politics has remained the same in substance, but so very different in outward form: no combination of the National Coal Board and landed gentry (assuming it were likely) could stand in the way of water power developments so easily today.

The Cooper Committee (1943), set up to investigate the whole question of hydro-electric power and economic development in Northern Scotland, was disturbed by the acrimony surrounding the subject, as a result of twenty years of political warring on the matter. It found 'grievance, suspicion, prejudice and embittered controversy' vitiating the promise of hydro power. The solution, adopted by a powerful wartime government, has worked well. The environment has been respected, new lochs have come to adorn the Highlands and turbine houses have been made neat and unobtrusive, not crass and ugly as the opponents of hydro power used to claim. The salmon and their owners have been treated with generosity, and employment has been created. None of these activities of the North of Scotland Hydro-Electric Board (the parliamentary creation which followed the Cooper Report) have sabotaged the Welsh economy or unduly hastened the decline of coal. However, had the environment-conscious not been so vigilant, the story might well have been different, for in a mass democracy, 'noise pays off'.

One last fact deserves comment: the relative decline in the power of the legislature. Nowadays, interest groups tend to 'pressure' the relevant government ministry. During the Electric Revolution they aimed more for Parliament, and notably the committees which looked at private electricity bills. The Acts of 1909 and 1919 had to be severely modified by the government to get them through Parliament, in the face of fast manoeuvring by pressure groups. Nowadays the executive is responsible for power matters *in toto* and would not hesitate to apply the party whips to get the required laws through Parliament.

In all these ways, the development of hydro-electricity foreshadowed much that followed. It gave the environmentalists and regional pressure groups some experience; it introduced another aspect of the 'rural *versus* urban' war that has gone on since the industrial revolution, if not before. The early electric visionaries prophesied that electric power would harmonize society and the economy; perhaps they would have been saddened to see that their vision could have proved so divisive, but that was the effect of grappling with reality.

NOTES

1 *Sussex Life*, Vol 6 No 5.

2 The Corwen system was of very respectable size, and at nationalization was the second largest non-statutory undertaking in Great Britain. There were two alternators sending out 3·3kV, stepped down to 220V for lighting and 380V for power. The dam was built by 'Colonel Wynn's own men' and held over one million cu ft of water; see *Electrical Review*, Vol 93, 1923.

3 The Giant's Causeway, Portrush and Bush Valley Railway & Tramway Co Ltd, see J. H. McGuigan's book on the system. It lasted 1883–1949.

4 18 October 1881.

5 It was not the only hydro-electric tramway, the Bessbrook & Newry Tramway (also in Ireland) was so powered; where coal was expensive this was a useful alternative. There were others, wholly or partly powered by this means, in Great Britain.

6 The first Welsh scheme of this type was at Ffestiniog (Yale Electric Power Co, 1900), where a more grandiose hydro-electric project is to be found today.

7 Finance came mainly from the Power & Traction Finance Corporation.

8 South of Scotland Electricity Board information.

9 MacColl was engineer on the Falls of Clyde Scheme, and remained with the North of Scotland Hydro-Electric Board until his death in 1951.

10 See Patrick MacGill: *Children of the Dead End*, which describes the scenes with eyewitness accuracy.

11 The scheme was run by the Lanarkshire Hydro-Electric Power Co, subsidiary of the Clyde Valley Electric Power Co, and financial help came from the P&TFC, see note 7.

12 Similar battles attended the Ebbw Vale Steelworks (which might have gone to Jarrow); and (post-war) the siting of the Royal Mint and the Post Office Savings Bank.

10, The Engineers

GENERALLY, engineers have wisely observed Goethe's maxim that life is to be lived, not cognized. Their achievements speak eloquently enough; a pleasing modesty and high professional standards have left their calling and their lives rather under-recorded. The electrical engineers had a profound effect upon the society in which they lived, and many of them had the unusual privilege of presiding over the whole Electric Revolution which they initiated or shaped. The revolution occurred within the lifetimes of Ferranti, Edison, Marconi and Crompton; by the time they had died electricity was being taken for granted. Even by 1910 Edison was treated as some kind of elder statesman. Merz, who was killed in an air raid in 1940, had his first major assignment at the age of twenty-four (1898), installing a lighting-and-tramways system in Cork; on his initial visit the jaunting-car driver delivered him at the gates of the poor-house, having no idea at all as to what a *power house* was! By the time of his death, Merz had seen some five thousand miles of national grid erected, and mammoth power stations like Battersea or his own Barking accepted as normal.

As in the industrial revolution, who one knew was almost as important as what one knew, especially when the demand for engineers who had mastered the techniques of the new supply system far exceeded their supply.[1] To come from a good 'training stable' or to know fellow engineers in the van of progress often put a man on the golden road. Thus, Parsons was trained by Armstrong, who knew Swan personally. Sir Alexander Kennedy taught Ferranti and also designed the Croydon power station upon which Merz worked in his first job. A bevy of Crompton's early assistants went on to be borough and city engineers while still young; Crompton knew Ferranti and counted among his close friends the wizard of engineering mathematics, John Hopkinson. Obviously, this would be a sensible training system for a fast exploding industry, especially in a country with very indifferent technical education facilities, as was Great Britain's case then. By 1920 the growth of formal technical training had largely replaced the personal contact system, although much remained, and a few vestiges still do.[2]

No work dealing with the history of electricity can realistically ignore the genius of Michael Faraday, who in his *annus mirabilis* of 1831 discovered the basic principles upon which electricity supply has rested ever since. But Faraday lived and flourished before the Electric Revolution; chronologically the first major figure of that phase is Sir Joseph Wilson Swan (1828–1914). Swan, with Parsons and Merz, was one of the Tyneside triumvirate that shaped the future of electric power supply. His major contribution was to invent the electric light, noted in Chapter 2 as the basis of electricity supply and engineering for many years. Swan was born in Sunderland, and spent his creative years in Low Fell, Gateshead and in Newcastle upon Tyne. Apprenticed to a druggist in 1842, he quickly blossomed into a prolifically inventive chemist, and contributed some important discoveries to photography and photo-engraving. He worried at the problem of producing an incandescent light from the 1840s (see p 7), had made carbon filaments for his light by 1855, and a short-lived burst of incandescence from a primitive bulb in 1860. In 1878 he produced the first true incandescent lamp, and later formed his company to exploit the matter. His last major innovation (found as a result of his search for a more perfect lamp filament) was artificial silk, from nitro-cellulose. In his later years he became the patriarch of electricity, with Neptune-beard, white locks and clear eye. The last honour bestowed upon him (posthumously, as it turned out) was the freedom of his native Newcastle, accorded at the same time to Parsons, inventor of the turbine.

Parsons, unlike his mainly bourgeois fellow engineers, was an aristocrat, being the younger son of an Anglo-Irish peer, the Earl of Rosse. Nothing could better show the haphazard state of British technological education in the mid-nineteenth century: Swan went to a Dame's School and thence to apprenticeship; Parsons received private tuition at the family seat (Birr Castle), but amid a background of considerable intellectual brilliance: his father was an astronomer of distinction. Parsons went on to Cambridge, emerging as eleventh wrangler in the mathematics examinations in 1877. Britain got better engineers than it deserved, and truly 'muddled through' the preliminaries to the great revolution ahead, saved by the genius of a few individuals.

Parsons' inherent love of mechanics led him to Tyneside, then as near to an engineering capital of the West as any city ever has been. He was a premium apprentice at Armstrong's[3] engineering works, Elswick. Later, working for Clarke, Chapman & Co of Gateshead, he produced his first turbine (1884), which was used as the engine of an efficient and compact lighting set for ships—the connection with electricity generation was thus established early in the life of the turbine. Going into business on his own account he developed the radial-flow turbine (1890), and shortly afterwards received a contract from Cambridge Corporation to light their streets from a turbine-powered station. Although the turbine later enjoyed great success in marine engineering, its greatest economic and social effect was in power generation, by making possible long-range AC distribution; DC turbo-dynamos were certainly made, but were much less common than turbo-alternators. Parsons was mechanically versatile, and a ceaseless innovator: he produced high-grade optical glass, and virtually founded the British high-grade optical glass industry; he dabbled also in acoustics and ballistics.[4] Like Swan, he lived to see his inventions widely accepted and fully integrated into the industrial and social framework; in 1931, the year of his death, it was calculated that the total capacity of all Parsons' turbines was 30 million horsepower. There were other turbines[5] but Parsons led the field by a large margin; he lived a quiet and fairly self-sufficient family life in rural Northumberland, a little cut off by his slight deafness. He was a sound business man, and benefited from his inventions, leaving an estate of £602,434.

Also of gentle birth, in this case of a Yorkshire gentry family, was Colonel R. E. B. Crompton (1845–1940), educated at Harrow, and starting off as an officer in the Rifle Brigade. Crompton made his name as an engineer by designing a massive steam tractor which hauled waggon trains about in British India, where he was stationed. He was awarded five hundred pounds for his services, and later decided to return to England and try his luck as a mechanical engineer (1875). Another branch of his family owned the Stanton Ironworks, Derbyshire, and persuaded him to light their works by electricity in 1878. With this sort of background experience, he carried out other contracts for arc lighting systems, for example at St Enoch Station, Glasgow. He also formed his own electric lighting company, the 'K&K'—(see p 85) in 1886. From the start he was a DC man, and can carry a great deal of the responsibility for starting the 'battle of the systems'. A fighter in other ways, too, he helped to form the Volunteer Corps of Electrical Engineers, which he took to the Boer War; and in the Great War he was one of the minds behind the invention of the tank. He died in harness, as a director of Crompton Parkinson, rich with honours (he was an honorary member of all three engineering Institutions) which ranged from a CB for his Boer War contributions, to the Crimea Medal (Sebastopol Clasp).[6] A man who had seen Sebastopol fall, and Battersea power station rise, he lived right through the Electric Revolution, and he died within eight years of its nationalization.

Unfortunately, by contrast, S. Z. de Ferranti (1864–1930) was cut off in his prime. Of all electrical giants, his education was the most cursory. As a schoolboy he played with various experiments described in the *Boys' Playbook of Science*, encouraged tolerantly by his mentor at St Augustine's Ramsgate, Abbot Egan. He learned about electricity mainly by working for Siemens, and attending a few evening classes at University College, London. Ferranti was half English, and altogether more wide ranging than other pillars of the electrical establishment—his Italian and generally cosmopolitan other half gave him an assured flair that made him the leading electrical propagandist of his day. One obituary of Ferranti[7] said that he was not an electrician of the calibre of Kelvin or Marconi: perhaps it ought to have said, of their *type*, for he had no equals as a proselytizer. His was the vision that inspired Deptford, and he was called its 'Michaelangelo', partly because of his Italian connexions,[8] and partly because 'from first to last, from foundation to top of highest tower, architecture, materials, foundations and machines, were all specified by one man'.[9] Ferranti was the marshal of AC in the battle of the systems, and in 1910 he delivered the historic speech to the Institution of Electrical Engineers (as President) which marks the culmination of electricity's 'spring'. He made uncompromising claims for total electrification, whereby all coal would go to make electricity, and no service would be powered by any other means. Whether this was immediately feasible was not the point: Ferranti had thrown down the challenge to coal and gas, and stated that electricity could power a modern society without help from any other power source. He threw his energy into many other schemes, for example the Electrical Association for Women (see p 138) and other 'missionary work . . . in the universal application of electricity for all the energy want of mankind'. He founded the great Ferranti engineering firm, took out 176 patents, and worked all his life at the problem of the gas-turbine which he felt to be the great prime mover of the future.

Of all the founders of the electrical economy and polity which we live in today, few are so little known, or so undeservedly, as Charles Merz (1874–1940). Although all the engineers mentioned were talented and excellent men, Merz was probably the finest man of them all. He had a horror of publicity, and maintained a strict professional silence—consequently he has remained a rather obscure figure to the history writers. His education was as excellent as Ferranti's was patchy: Bootham School, and Armstrong College, Newcastle. He came from a deeply religious and intellectual family[10] of Newcastle. The first impression that Merz makes is the ability he had to accept weighty responsibility when young: at 24 he electrified the City of Cork; by the time he was thirty he had installed the pioneer three-phase electrification and established a reputation as a leading parliamentary witness: 'a superb expert witness, calm, collected, lucid, accurate and penetrating'.[11] He believed in electricity as strongly as Ferranti, although rather less publicly—even in 1904 it was said of him that 'if he were asked he would electrify the whole of England, with a branch line to America'.[12] He founded the consulting engineers, Merz and McLellan and designed or guided hosts of important pioneer schemes: North Tees and Barking 'super power stations', the Tyneside and Melbourne railway electrifications, much of the national grid, the North East coast integration, etc. He advised the government on many occasions, but would accept neither payment nor honour for this, which he saw as straightforward duty.[13] Ironically, this pacific Quaker was killed with all his family (except Mrs Merz) in an air raid in 1940. Like the similar death of Lord Stamp this was a great national loss at the worst possible time. To quote Wren's epitaph: *si monumentum requiris, circumspice* has become rather a cliché, but it is easily the most fitting monument to a man who did more to shape modern Britain than many a second-rank politician who is to be found adorning the pages of history textbooks.

In addition to these rather arbitrarily chosen leaders of engineering there have been legions of competent and essential men of whom much less is known or recorded. Such men are: Sir Edward MacColl, who gave much of Scotland hydro-electric power, and William McLellan, Merz's partner who drove himself to an early death fighting for the Galloway scheme.[14] There were the electrical academics, like E. T. Dover who trained generations of traction engineers, Sir Alexander Kennedy, and John Hopkinson. John Hopkinson (1849–1898) is a fine example of the engineer-academic who watched over the revolution that he was helping to create. He possessed a painstaking and brilliant intellect, and was foremost a scientist, then an engineer. He bears intellectual comparison with Nikola Tesla, although he was very different as a man, being a methodical Nonconformist family-man of the most impeccable English kind. His pioneer work penetrated many fields: lighthouse illumination, magnetic circuits (really his invention), traction, generation and distribution of power—the multiple wire system of distribution (see p 75) was one of his brainchildren (1882). He worked well in a team, and was one of nature's teachers; he was appointed to be first incumbent of the Chair of Electrical Engineering at King's College, London University (1890), and he was twice President of the Institution of Electrical Engineers. From Hopkinson there poured a seemingly limitless stream of patents, papers and soundly trained minds. His untimely death in an Alpine climbing accident was a tragic loss both to electrical engineering and his country.

There were other memorable teachers of electrical science and engineering, whose own researches and didactic talents spread the gospel of the Electric Revolution. There were for example, Professors W. E. Ayrton, John Perry and Sylvanus Thompson. The former two operated at the City and Guilds Institute, Finsbury, and were doubly distinguished in that they were among the first teachers of electrical engineering in Japan, during the 1870s. Thompson (1851–1916) was one of the father-figures of the Electric Revolution, author of many and voluminous standard works, and later the biographer of Lord Kelvin, himself a scientific polymath with a fervent belief in the efficacy of electric power, and the most prestigious member of the Direct Current school in the 'Battle of the Systems' (see Chapter 7). Mere lists of past engineers would be not only tedious, but unjustly selective. More important, perhaps, is the record of their professional and trade associations which made them into a respected and powerful class of men.

In the early years of the Electric Revolution, professional and other associations were very perfunctory. Engineers kept in touch by actually knowing each other, by reading the technical press[15] and occasionally exchanging views at meetings or dinners, like Crompton's 'Dynamicables'. So quick was the Electric Revolution that more formal and continuous contacts became necessary, both to protect interests, and to guarantee to the public certain standards of competence and experience. On the lines of the Institution of Civil Engineers (1818, chartered 1837) the Society of Telegraph Engineers was founded (1871) for 'facilitating the exchange of information and ideas among its members'. The Telegraph Engineers represented the largest single body of 'applied electricians' at the time, but as other applications for the new power grew (as well as the fact of its actual generation) so, in the words of a member, C. F. Varley (1872): 'The Society will gradually, by natural selection, develop more into an electrical society than into a society of telegraphy proper.' Varley's darwinist prophesy was borne out when the Society added 'and Electricians' to its title (1880) and in 1887 when it became the Institution of Electrical Engineers, which it has remained ever since. Its membership grew with the electrical age: 155 in 1873, 1,500 in 1890. In accordance with its opening manifesto, it spread knowledge and encouraged electrical education by collecting a superb library and agitating for wider technical education, and it also published its own *Journal* and *Proceedings*. As far as customers and the informed general public was concerned the main achievement of the Institution of Electrical Engineers was to elevate the reputation of electrical engineering by its insistence on high standards of conduct and training as a prerequisite of membership; it also established a 'hierarchy of experience' whereby senior grades (such as MIEE) were only open to engineers of proven high ability.

Among the many events that mark the end of the Electric Revolution in the early 1930s, was the publication of Rollo Appleyard's *History of the Institution of Electrical Engineers, 1870–1931*. It was akin to that gesture of the Emperors of China, who marked the arrival of a new dynasty by ordering 'that the history of the previous dynasty be written'. Appleyard's book, both in its title and content was conscious that the pioneering days were over, and he wrote *finis* to them.

There were many other important organizations which emerged out of the fluid changes of the Electric Revolution, and became hardened and established. The fast-growing body of workers who kept the power houses operating, laid the mains and serviced the meters were organized into the Electrical Trades Union (EETU/PTU since 1968), which was an industrial union born when the concept of industry-wide unions was in its infancy. Unfortunately, while so much work has been done on the macro-organization of this and other industries, there remains much that is unknown, and always likely to be, about the origins, names and sources of so many early electrical workers. Some came from agriculture, then (1873 onwards) smarting under the blows of foreign competition; others came from Ireland and others from industries which offered inferior wages, security or satisfaction. A great wealth of industrial folklore and history has been sadly lost.

However, the workers' own organizations are well recorded. Like the Institution of Electrical Engineers, the Electrical Trades Union grew out of the original largest bloc of specialists, the telegraph men. It was formed in 1889 out of two small local unions[16] and numbered 570 members. It was open to wiremen and linesmen, but also (bearing in mind its aspirations to be an industrial union) 'labourers of six months standing'. It levied sixpence (2½p) per week per member, flat rate, and worked for the normal trade union and friendly society objectives, ie 'for resisting any attempt to curtail or take away any of those privileges which are or may become the customs of the trade'. The Electrical Trades Union published the *Eltradion* (later *Electrical Trades Journal*) for its members, and in 1907 moved its headquarters to Manchester where there was a large concentration of members. At first membership was erratic, but the Edwardian inflation, falling real wages and some hard-line militancy by its leaders swelled its ranks, so that whereas it only had 1,871 members in 1910, this had risen to 8,195 in 1914. In 1919 it was invited to sit on the 'Whitley Council' (National Joint Industrial Council) for the electricity industry.[17] The ETU's early organization, quick growth and acceptance all mark it as another early-maturing part of the Electric Revolution.

The suppliers of electrical goods and electric power were themselves organized before the Great War as well. The municipal IMEA (Incorporated Municipal Electrical Association), 1895, and the private IAEPC (Incorporated Association of Electric Power Companies), 1905,[18] existed for mutual protection, and to pressure public and official opinion; the IAEPC did add also, as one of its official objectives 'to encourage, expedite and facilitate the development of the electric power industry'. Manufacturers also had their associations, like the Cable Makers Association (1899), and the Electrical Contractors' Association (1901). The all-embracing BEAMA (British Electrical and Allied Manufacturers' Association) dates originally from 1903, although in its present form from 1911. It represented the electrical engineering industry as a whole, and has always carried great weight in negotiations—it was one of the more effective advocates of standardization of supply, for example. It could count among its Vice-Presidents (1925) such men as Lord Weir, Ferranti, Parsons and Callender, of Callender's cables. Its later established majesty has obscured the rather panicky nature of its birth as the National Electrical Manufacturers Asociation (1903) when it was a hastily mobilized defence group of British manufacturers trying to stop the onslaught of foreign competition. These were the years when Westinghouse and American GEC were setting up establishments in Great Britain, and also when Free Trade was a widely held dogma.[19] For many years the guiding spirit of BEAMA was D. N. Dunlop, one of the great electric propagandists, with some very long-range plans for the industry. It was he who started the World Power Conferences',[20] for example, in order to meet these long-term problems with prepared plans, rather than hasty and makeshift tactics.

Many large firms and institutions found, by 1914, that they had to have an electrical engineer on the staff, and a body of technically trained subordinates to help him. The Army and Royal Navy responded to the need very early in the day: the Royal Engineers Corps of Electrical Engineers (late REME) was founded in the 1890s, and detachments fought in the South African War 1899–1902.[21]

One factor is common to all these groups and associations: they were established before 1914, many before 1905. Yet they would not have been necessary before 1880. In the intervening twenty-five years the Electric Revolution had got off the ground, and was well aloft: no great industry had grown into such size, complexity and establishment in so short a time before; this alone would justify the term 'revolution'.

NOTES

1 A typical such figure of the industrial revolution was Henry Maudslay who trained Nasmyth and Whitworth, and himself studied under Bramah.

2 For example, to be a 'Metrovick' apprentice was long considered to be a great privilege, and the qualification was well worth having as late as the 1950s.
3 Another mechanical wizard without formal training: Armstrong started as a Newcastle Solicitor, with a *taste* for hydraulic engineering.
4 He invented the first rifle silencer. Because he was slightly deaf, it is claimed, he welcomed good amplification, and he contributed to this the 'auxetophone', a huge trumpet-affair which amplified stringed instruments very successfully—Sir Henry Wood praised its high fidelity. On a lighter note, Parsons was the only man who could abide the terrible screeching of his sulphur-crested cockatoo pet, a compensation however slight for his deafness.
5 Notably the Ljungström with contra-rotating blades, and the Curtis. A few of the latter were built vertically, looking rather like the alternators in a hydro-electric plant: Liverpool had some at Lister Road No 2 station, and the Yorkshire Electric Power Co at Thornhill.
6 He did not fight in the Crimea, but was present as a schoolboy observer.
7 By Ll. B. Atkinson, MIEE, *Electrical Review* 1930 .
8 He was named after a remote ancestor, Sebastiano Ziani, 41st Doge of Venice (1172).
9 See Gertrude de Ferranti and Richard Ince: *Life & Letters of S. Z. de Ferranti.*
10 His father, Dr Theodore Merz, was a director of British Thomson-Houston and NESCO, also author of *A History of European Thought in the Nineteenth Century*, 2 Vols.
11 For Merz as a witness, see *Dictionary of National Biography.*
12 Said in connexion with the proposed London rationalization, see p 57.
13 He was thrice-pressed to be President of the IEE, the highest professional honour; he declined because of his dislike of public speaking, which would have been required of him.
14 He was found dead one day, lying beside the path from his rented cottage to the Galloway workings.
15 See p 139.
16 The Amalgamated Society of Telegraph and Telephone Construction Workers (Manchester) and the Union of Electrical Operatives (London).
17 Other unions sit on the NJIC as well as the EETU/PTU; the AEF, the TGWU, the NUGMW; the CAWU, the EPEA, and NALGO. All of these except NALGO are affiliated to the Confederation of Shipbuilding and Engineering Unions, which covers the electrical engineering

industry as well. Their full titles are: EETU/PTU: Electrical, Electronic and Telecommunication Union/Plumbing Trades Union. AEF Amalgamated Union of Engineers and Foundry Workers. NUGMW: National Union of General and Municipal Workers. CAWU: Clerical and Administrative Workers' Union. TGWU: Transport and General Workers' Union. NALGO: the National Association of Local Government Officers. EPEA: Electric Power Engineers' Union.

18 For other comments on these, see p 45.

19 Free Trade was under fire at the time, and was shortly to damage the Conservative Party in the 1905 General Election, when one part argued for protectionism. In the event Free Trade lasted, with a few exceptions, until 1932.

20 The first was held in 1924.

21 The Royal Navy had no separate electrical branch; electrical engineers came under the Engineering Branch, but in 1925 an 'electrical category' was started, and Electrical Officers (who were originally warrant officers) staffed it. For many years they came under Torpedo Officers, but have come to constitute a separate branch now.

11, Overseas

'We are living in the dawn of an electric age', every stream in America tumbling 'from the mountains through the meadows to the sea', a potent source of power 'from the snows, the springs and the rain'. He saw America as a shining land of turbines, generators and transmission lines, white dams brilliant in the sun with sparkling blue water behind, power flowing to the people at a price that all could afford.

Arthur M. Schlesinger, Jnr quoting and describing Senator George W. Norris.

THE course of the Electric Revolution beyond Great Britain throws Britain's own problems and developments into relief. Any number of examples would do this, here the matter principal examples selected are the USA, Ireland and the USSR, with further matter drawn from contrasting areas. Each country created its own peculiar problems, and worked out its own answers to them. After studying these matters one is forced to conclude that Great Britain's tribulations were often no worse than anyone else's, and were in some ways better handled. One element becomes indisputably clear when the Electric Revolution is viewed 'globally'—it was, as the British experience showed—a supremely political matter. So great an economic and technical matter was almost certain to become such in an age when politics has increasingly come to mean the resolution of economic problems.

The USA was as much an electrical pioneer as Great Britain or Germany. By virtue of its large and fast-growing population, its economic strength and very newness, the USA was an obvious choice for electrical development. Many factors favoured it:

cheap hydro-electric power, abundant coal, densely settled industrial areas, a numerous and modern-minded bourgeoisie, energetic businessmen and talented engineers. The country of Edison, Niagara, Westinghouse and the Interurbans! By 1900, the USA was probably the world leader in electrical development, followed by Germany. It exported its techniques and capital to Great Britain, thereby for the first significant time turning the largest creditor-nation into a partial debtor. This contrary trend in Britain's capital balance was relatively small, but the heavy investment represented by British Thomson-Houston (ie American GEC) and British Westinghouse was a sign that Britain was beginning to slip, and 'Yankee hustle' as it was called was now setting the pace in new industries.

By 1921 there were nearly seven million American homes electrified, and they were served by 5,600 electricity supply companies, or municipalities. In 1921 forty-two billion kWh[1] were sold to American consumers, for just over a billion dollars. From 1910–20 the output of generating stations trebled, and demand was both cultivated and sustained by the most energetic and sophisticated 'hard sell' of any industry's marketing. British attempts at marketing (criticized by Weir in 1925) were feeble and jejeune by comparison, although in some ways this enhanced the quality of life, at least for those who resented the lurid publicity and effervescent badgering of the salesmen. However, for all that, America had then, as now, no peer at marketing, and the electricity industry benefited from this.[2] About half of America's power was water-generated, and already big regional networks were being linked up by grids of what was called 'super power': high tension transmission of up to 220kV. The four main power reticules were: the Gulf States-Tennessee; New England; Pennsylvania and California. In addition to the seven million electrically lit, and sometimes electrically powered homes, there were a further 326,840 industrial and commercial users of power, with the heaviest concentration in the North-Central region.

With such opulent and regionally compact markets, it is not surprising that America could boast large and efficient power stations, and, in the absence of the type of restrictive legislation suffered in Great Britain, large power-generating companies. In 1920 there were seventy-one companies generating more than one hundred million kWh each, and nine companies generating a billion kWh each. The three largest companies were: the Niagara Falls (NY) Power Company; the Commonwealth Edison Co of Chicago and the Pacific Gas and Electric Co. Municipal ownership was not so advanced as in Great Britain, and had recently suffered a severe set-back during the Great War inflation period, when several installations were either sold off to private companies, or connected to private grids (thereby shutting the town generating plant) in order to ease the city budget which was under pressure from mounting costs.

On the surface, the year 1920, lying as it did on the brink of the years of Coolidge prosperity, seems to present a picture of organized and buoyant prosperity in the electricity supply industry. In some ways it did, but all this consumer-led urban affluence obscured some deep ills in the industry. These were to intensify in the 1920s and become a major scandal and political issue within the decade. In this way, the electricity question became a principal target for the populist reform tradition, and a large element in F. D. Roosevelt's New Deal programme of economic and social reform. The industry was to receive the sharp and effective attention of radical legislators, and it was both tamed and transformed in the process.

There were two main issues at stake. First, rural America (which was not sharing the national economic boom, in any case) was very patchily electrified, and many areas had no electric power at all. They remained, however, politically powerful. By 1932 only one farmer in one hundred had electricity in Mississippi, and one in twenty-five in Tennessee or Alabama; nine-tenths of American farms used fossil fuels or animal power. The power company philosophy was urban-oriented and uncompromising as far as rural areas were concerned, it was simply the dictum: 'low consumption, high rates', and these rates being high left a large market untapped. Rural America could not afford power in these circumstances, and the city-based power companies were very sceptical of building transmission lines over large distances to lightly populated farming regions. As rural America slipped behind in the economic league, its chances of electrification seemed to get slimmer. There were politicians who took up the issue, conscious of a mounting injustice, and occasionally of the big political capital to be exploited in the matter. The leading, and genuine, electrical radical was Senator George W. Norris, of Nebraska. Under Norris's influence Nebraska became (and remains) the only State in the Union to go into the business of large-scale electricity generation and distribution. According to Norris: 'This [electricity] is a property that belongs to all of us—a source of human happiness.' This is the authentic voice of the Electric Revolution, it echoes Ferranti; together with the quotation employed as a chapter heading (see p 118) it serves well to illustrate the strong passions engendered in America by this problem. John Dewey, the philosopher, did not exaggerate when he said: 'The power issue is the most weighty single issue in the political field.' As a United States Senator, Norris was to take the lead in fighting the power companies, and widening public supervision of them.

The other problem lay in the nature of the power companies themselves. The electricity supply companies were very often organized into immense and elaborate combines, so carefully constructed that they defied standard Federal anti-trust legislation with impunity. They were monuments to Faustian finance, and examples of how capitalism could organize in its own, rather than the public, interest if 'let alone' according to the classical doctrines of Old Liberalism. The message had been clearly read across the Atlantic, and accounts for the great hostility towards 'power groups' expressed in Great Britain, particularly by municipalities (see p 42). There were three main groups, organized as delicate pyramids of holding companies, whereby relatively small holdings of equity stock in key companies enabled the holder to own and influence scores of dependent companies. The 'utility groups' (which often combined gas and water supplies with electric power and traction in their portfolios) were: the United Corporation; Electric Bond and Share, and the Insull Group. By 1932, the last year of the pre-Roosevelt era, these three generated 40 per cent of all electric power in the USA; thirteen holding companies generated three-quarters of all privately produced power there. The system came under fire for many reasons. It offended the 'new' liberals, the populist tradition and the anti-trust interests by its very nature; it was said to charge unduly high rates for power; it was believed (often with cause) to employ graft in order to win favourable local franchises. The manipulators of holding companies expended most of their energy and talents in financial legerdemain, not in rationalizing output—and, as noted, rural America was largely bereft of electricity, and the sworn foe of the power combines. Efforts at regulating these companies were not very successful: municipal politics were too small (and softened by corruption, on occasion) to handle the matter. State legislation, and supervision by public service commissions made some useful if modest headway from *c* 1910, and the Federal Power Commission was only set up in 1920, to supervise interstate electrical matters. None of these institutions had proved sufficient match for the power and craft of the holding companies, and their dark labyrinths of finance and organization. All of this took place, of course, in an era when the confident and optimistic business mind held a firm grip on public opinion. Many Americans openly admired the financial dexterity of such past-masters of holding company mystique as Samuel Insull .

It was the collapse of Insull's mid-Western holding empire that called the system into question with majority opinion. Insull's group (containing *inter alia* such giants as Middle West Utilities and the Utility Securities Co) had absorbed millions of dollars of investment from the public during the great Wall Street boom of the mid-1920s. In eight months before the 'Great Crash' of October, 1929, Insull Utility Investment shares rose from thirty dollars to one hundred and forty-seven dollars; the whole group was capitalized at about two and a half billion dollars. After October 1929, the entire edifice disintegrated; millions of dollars were lost by investors and investigations showed up extensive and complicated irregularities, corruption and dishonest management. Insull was indicted on various charges, and although he was acquitted, irreparable damage had been done to the holding company image. Insull had been highly unpopular with the radicals during his gilded years, now he was detested most heartily by masses of small investors, who hitherto had been neutral or trusting.

The New Deal era lies rather beyond Great Britain's Electric Revolution, but its effects on electricity supply are an important comparison to note. Many of F. D. Roosevelt's supporters were sworn enemies of the electrical *ancien régime*. Their policies broke the holding companies, and electrified rural America. Their symbolic masterpiece was the Tennessee Valley Authority (TVA), which was to become the very embodiment of New Dealism; one of America's rare excursions into nationalized (or, as preferred there, 'socialized') industry. The TVA was a vast project of rural improvement, area drainage and electricity generation. It was the large-scale extension of a more modest scheme (the Muscle Shoals project) launched in Woodrow Wilson's presidency, and which narrowly survived an onslaught by President Hoover whose dogmatic belief in the virtues of private enterprise were upset by this small exception. Fortunately for the future TVA, Hoover was outmanoeuvred by Norris. As well as being a most controversial project in itself, the TVA suffered internal wars in its early years. Broadly speaking, the debate raged around its relationships with nearby private power companies, notably the Commonwealth and

Southern, whose chief (Wendell Willkie) strongly opposed the TVA, and what he supposed was its subsidized, socialistic competition. The TVA chief, A. E. Morgan, was inclined to compromise; his lieutenants, and in particular David Lilienthal, were not. They had an altogether more rigorous and idealistic view of the TVA's functions, and their school of thought won. Thenceforth, the TVA became one of the most successful examples of state enterprise anywhere.

At the same time, another New Deal agency, the Rural Electrification Administration (1935) pressed on with the electrification of rural America, so long delayed. Its first chairman, Morris Llewellyn Cooke[3] was an old campaigner for rural electrification. He found out, by trial and error, that the best means for the REA to achieve its ends was to encourage the formation of local electric 'co-operatives', set up with Federal help. The existing utility companies were unwilling to help, and municipalities were similarly unable or unwilling to do the business with any despatch. So Cooke, and his successors, helped the countryside to help itself, and by 1950 nine-tenths of American farms had electricity: the reverse of the situation twenty years before. Arthur M. Schlesinger, Jnr, rightly remarks of the REA: 'no single public agency ever so enriched and brightened the quality of rural living'.[4] The last battle was with the utility groups, which were brought to heel by the Public Utilities Holding Companies Act (1935). They were forced to divest themselves of many subsidiaries (a major blow to the already very sick interurban electric railways) and it was made statutorily impossible to erect 'Insull pyramids' ever again.

By ensuring the strictest supervision of company finance, the Securities and Exchange Commission prevented any holding company conjuring in the future. By 1963 the eleven holding companies that remained supplied 21 per cent of privately generated power, itself about three-quarters of America's total. Many of the remaining private companies amalgamated properly in order to enjoy economies of scale: four thousand companies were reduced to 480. These integrated companies were in turn carefully supervised by State public service commissions.[5] Interstate rates and services are further regulated by the Federal Power Commission. Thus the USA enjoys a varied 'mixed economy' in power matters, closely supervised in the public interest. A body of critical opinion continues to press for total publicly-owned power generation, and it is vigorously opposed by the companies. Superficially, this looks like pre-Weir Great Britain, but owing to the great size of the USA, and its regional variations, the system is better suited to that country than the old muddle that afflicted Great Britain (a far smaller country) during the Electric Revolution. By 1962 the American power pattern had settled down to this pattern:

3,617 systems, of which
(*a*) 480 were private, with 76% of generating capacity, serving 79% of the retail customers
(*b*) 2,124 state, or municipal, 10% capacity, 13·5% of customers
(*c*) 969 Co-operatives; 1% capacity, 7·5% of customers
(*d*) 44 Federal (including TVA); 13% of generating capacity.

The electricity problem in Ireland presents an interesting variation of the British one, of which (in a sense) it was a part until 1922. Northern Ireland and the Irish Free State (now the Republic of Ireland) dealt with the matter in different ways; the latter is the case in question for this chapter. Up to 1922 Ireland operated its public utilities on the British system, since it was statutorily part of the 'United Kingdom of Great Britain and Ireland' in those days. The Irish Free State started, therefore, with a typical pre-Weir patchwork of municipal and private company undertakings. The whole pattern was heavily urban-dominated: Dublin took three-quarters of all electricity produced (forty-eight million units sold in 1922). Rural Ireland, like rural England and America, was under-electrified by later standards. Ireland had another problem which did not affect Great Britain or America: it had to import virtually all of its power-generating coal. These matters presented a strong challenge to the leaders of the new state, and they responded with vigour. The most important action was to accept the far-seeing proposals of Dr Thomas A. McLaughlin and Siemens-Schuckert (the German electrical engineers for whom he worked) to harness the river Shannon in order to generate hydro-electricity. At a cost of five and a half million pounds, it was hoped that the Irish Free State could free itself from the need to import so much coal. The scheme was carried out from 1925–9, at Ardnacrusha—the optimum point on the Shannon for constructing a generating station (see figure 17); at the end of a nine-mile loop which gave a hundred foot fall of water. Some 252 million cu ft of earth and thirty-nine million cu ft of rock had to be excavated before the scheme was finished, and handed over to the new Electricity Supply Board. In later years this Board bought out the independent undertakers, and the last one went in 1968. Ireland also developed turf-fired power stations, and, not having a declining coal industry to support, was rather more bold in building oil-fired ones than Great Britain.

British interests were rather put out by this incursion of German technology into what they had always taken for granted as their sphere of influence; but the newly independent Irish understandably wanted value for money first and foremost, and this Siemens-Schuckert could offer. The Shannon scheme, and the gradualist, pragmatic change in Irish power matters stand in clear distinction to the political dogmatism and excitement surrounding the power industry in (then) contemporary Great Britain or the United States.

Electricity was to supply another political goal in the 1920s—in Soviet Russia. In the first bloom of zeal following the Bolshevik Revolution (1917) two enthusiastic apostles of technology drew up plans for mass-electrification in Russia: I. G. Aleksandrov and V. I. Grinevetsky, whose 'Goelro' (based on the Russian acronym from 'State Plan for the Electrification of Russia') called for large regional power stations, and an output of 8,800 million kWh within ten to fifteen years (Cf 1,900 million kWh in 1913). Lenin liked the idea, and saw its immense political appeal, hence his classic dictum (in full): 'Communism is Soviet power plus the electrification of the whole country';[6] he ordered the embodiment of Goelro without delay. Goelro was overtaken by Stalin's first five-year plan (1928), but its spirit and purpose remained intact, and by 1940 the USSR's output was 48,300 million kWh. The prevailing symbol of Soviet electrification in those years was the crash development of hydro-electric power, especially popular in the late-Stalinist era. The programme was severely modified after 1958, when it was realised that other interests like agriculture and fishing were insufficiently co-ordinated with hydro-electric schemes. Both Lenin and Stalin fully realized the paramount importance of electrifying the Soviet Union, if it was to be economically and strategically strong. Opposition to their gigantic programmes would have been neither economically sound, nor politically wise. Thus Krzhizhanovsky (first in charge of carrying out Goelro) and his successors did not have to cope with the kind of opposition that faced MacColl and McLellan, or the TVA and the Grampian Electric Power Company.[7]

Each State developed its own style and answer to the economic and political problems surrounding the need to electrify. In France a vigorous and well-managed finance group, the Messine Group (controlled by La Lyonnaise des eaux) acted as the great electric consolidator and entrepreneur during the 1920s. First it rationalized part of the Paris supply by forming L'Union d'Electricité (1919), and then erected a central station at Gennevilliers (1922). Many parts of the British Empire and Commonwealth copied the mother country's basic electric law. The *Electrical Review* of 1912 mentioned Queensland, Victoria, South Australia and Western Australia as well as British Guiana and Jamaica, as having statutes or regulations influenced by the Electric Lighting Acts. Canadian provincial and federal government was less inhibiting, and Ontario (1906) initiated large-scale state organization of hydro-electric power. Italy, under Mussolini, ran a curious compromise system whereby the state liked to have one key power station in each 'power region', but allowed private ones to co-exist. The main force behind Italian electrification was to save imported fuel, like Ireland. Before the Great War, much of Italy's railway system had run on British coal. The widespread electrification of Northern Italy's railways on the three-phase system from 1900, but especially from 1920 was in answer to this problem of imported fuel. As already remarked, this electrification had as much to do with 'making the trains run on time' as ever Mussolini did—*Il Duce* was the beneficiary of the fixed-speed characteristics of the three-phase motor! Near Genoa, for example, three-phase AC (at $16\frac{2}{3}$ Hz) was generated by the State; by the State Railways and by SFIM (Societa forze idrauliche della Maira) in 1924—a compromise which presumably reflected the political realities of Fascist Italy.

It would be tedious to list electrical development, state-by-state, proving ever more obviously that the generation and supply of electricity is both a political and an economic matter. But before leaving this international survey, it is worth noting how far the Electric Revolution had gone in the 1920–30 decade. In America, electricity was being transmitted on 'super-power' grids at up to 220kV by 1923; Japan and Germany also had them, and the Rhenische Westfalische Elektrizitätswerke led Europe in this development by the mid-twenties. Even in 1920, Italy, India, France, Spain and Sweden all had 100kV lines: significantly, Great Britain did not share in this until the coming of the National Grid. By 1930 'Ontario Hydro' plant at Chippewa had a capacity of half a million horse power, and each turbine of the Niagara Falls Power Co works developed seventy thousand horse power; one year later the Dniepr hydro-electric plant was ready in the USSR (capacity: 558,000kW). In Germany the Allgemeine Elektrizitäts Gesellschaft had just finished the first really large diesel-electric station at Cottbus, and what was Europe's finest super-power station, the Klingenburg station, near Berlin: integrated, spacious and functionally attractive, it contributed about 300,000kW to the (then) estimated total output of the developed world, forty million kW. A suitable note to close this global survey! By 1930, as Britain's Electric Revolution ended, men were speaking of so many millions of kilowatts being generated. No one has accurately recorded or estimated the total world output of electricity in 1880, but it was obviously modest, even tiny, by comparison. Who could expect otherwise in a world in which this was uttered: 'It has been proposed of late to supply electromagnetic machines in lighting streets and workshops, and the experiment has been tried with some success.' (*Encyclopaedia Britannica*, 9th edition, 1879.)

Within twenty-five years of this cautious recognition, the electricity industry was strongly established in many parts of the world; governments were setting about regulating it, and in some cases were beginning to invest in it themselves, for which reason the Hydro-Electric Power Commission of Ontario had great seminal significance. The idea of democratically-appointed commissioners ensuring an abundant supply of cheap power for the people seemed unexceptionable to the liberal mind. All over the world, electricity seemed a liberalizing force, offering the masses a higher standard of living. A. J. P. Taylor, writing of inter-war Europe in *Sarajevo to Potsdam* thought: 'Civilized life was made easier and brighter. It was also made more democratic. Electricity was an economic proposition only if everyone, or nearly everyone, used it. Soon the humblest homes were lit more brightly than the grandest palaces of previous ages'. However, there was another side to all this. Power generation and distribution was either organized, or supervised (or both) by state authority. As societies became more or less dependent on electricity, so governments acquired new and vast areas of influence. Of course, this does not necessarily imply some sort of 'electric despotism' (Cf

Karl Wittfogel's hydraulic basis of *Oriental Despotism*, in his work of that name)—but it does extend the dependence of the citizen on his state most dramatically, and to a degree that could not have been guessed in 1880. Nor is there anything wrong in the process; but it must be remembered that, as Lloyd George shrewdly noticed, electricity is as much a matter of politics as of engineering.

NOTES

1 A billion here being an American billion, ie one thousand million. One regrets the decline in the use of the very useful European *milliard* for the same number.

2 Salesmanship had been too successful in the Great War, and the US Government had, *inter alia* curbed the sales of electric ovens, and put an embargo on the use of electric display signs; Cf Great Britain when faced with 'Winter emergencies' of the power go-slow in 1970.

3 He had been connected with the Pennsylvania 'Giant Power Survey' of 1923, initiated by Governor Gifford Pinchot by way of encouraging rural electrification in his State.

4 Arthur M. Schlesinger, Jnr in *The Age of Roosevelt*, Vol 3.

5 The standardization of power in the USA on AC at 60 Hz was effected entirely by co-operation, not by Federal or State intervention.

6 Often rendered as: 'Communism is Socialism plus electricity', see p 4.

7 For the interesting connexion between American and Russian hydro-electric methods and philosophy, see W. H. G. Armytage, *The Rise of the Technocrats* (1965).

Part Three

12, An All-Electric World

BY 1910 the electricity industry was formally organized, and conscious of its important status. It looked forward confidently to an electrically-powered future civilization, which already seemed a real possibility. Yet it remained, in many ways, young at heart. It said and did things in the way of the *parvenu*, and of a type that the well-established railways would have avoided with distaste.

An example of this innocence appeared in *Simple Stories*, published by the *Electrical Review*.[1] The stories were written in a light satirical vein by 'Innocens', and usually described actual contemporary incidents. One for example *(L'Enfer)* told how a boatman unwittingly rammed an iron mooring bar through the towpath and into electricity mains—this had actually happened in London. The implication was that the boatman thought that he had poked through the lid of Hell, and the story proudly and implicitly assumed that electricity was the mightiest of slumbering giants, able to surprise and devastate when disturbed. It was written in pseudo-romantic style, opening: 'It is night. The moon swims deliciously in the billowy cloudlets, like a nymph disporting herself in the caresses of the ocean.' Later, it facetiously described the bargee's landing: 'What grace! What agility! *Ma foi* what a man. He thinks of wife and family. . .' Poor innocent, for 'under the towing path—ah, the perfidiousness, the subtelty of electricity, my cherished ones—lie the cables electriques of Villesdenne'.[2] After seeing his bar 'volatilized' the boatman jumps into the canal, and in a note of jarring bathos the story concludes: 'In the distance the engines of the electric light station have shut down, for it is of no utility to them to run on a dead short consisting of a bargee's bar.'

An incident in the Middlesbrough electric lighting committee, whereby the borough power station costs included a charge for 'milk' was, naturally, a great temptation for Innocens. The Teesside borough became Centreburgh, and a satire on the bureaucracy and intensity of municipal public life duly followed. The basic problem, it transpired, was mice; should the volt boy[3] have at them with a bamboo rod? Far better to employ a mouser with low initial cost (the Borough Electrical Engineer could 'get the beast for nothing') and it would remove the mice from the busbars, where their tails were causing shorts. This was well worth the expense in cat milk '0·0000067d per unit delivered at consumers' terminals'. A compromise ended the matter: give the cat less milk, and so encourage its zeal.

The irrepressible *Electrical Magazine* expressed electricity's youth best of all; it was not so comprehensive as the more measured *Electrical Review*, or *Electrical Times*,[4] but it made up for that with its buoyancy. Imagine the explosion when the *EM* noticed a cartoon in *Gas World* (1905) showing a fine sailing ship (the *Gas*) with a long oriflamme pennant 'light heat and power' moving gracefully through a calm sunset—the setting sun being marked: 'Electric light waning'. 'A certain well-known contemporary, whose contents are perforce "gassy",' was treated to a reply cartoon. The good ship *Gas* had been struck by some very electrical looking lighting, the zig-zag bursts were marked 'Light, heat and power' and emanated from a Statue of Liberty-esque beauty of chaste and forbidding whiteness; *Ships that pass in the night* was the contemptuous caption. Later generations, fancying themselves to be more sophisticated, might find such advertising rather puerile, but on the other hand we might envy the breezy confidence that such art seems to manifest.

The missionary fervour of the early electric propagandists

found many outlets. The *Electrical Magazine* launched a special supplement (1906) called *Popular Electricity*, which spoke of: 'A Great Revolution . . . sweeping over industry, commerce and domestic life; it has crept forward . . . through the outposts of the past[5] outmanoeuvred prejudice and overrun precedent . . . electricity has come out on top by sheer weight of merit alone.' This feeling of the pure logic and justice of the electrical case was implicit in a famous and seminal speech which Ferranti made as President of the Institution of Electrical Engineers, in 1910.

After preliminaries, Ferranti launched into one of the most prescient and comprehensive technical addresses of the age: 'I propose to address you this evening on the following subjects: coal conservation, home-grown food and the better utilization of labour.' He spoke at length, but with a terse and matter-of-fact style. Broadly speaking he advocated nothing less than an All-Electric economy, and indeed the phrase, if not new, was virtually launched that day. All coal was to go to generate electric power, all processes were to be adapted to it.[6] This would cost five hundred million pounds, and use sixty million tons of coal per year. He outlined one hundred super-power stations, each of ten 25,000kW generators to do the job (in addition to existing machinery, that is). He believed that gasified coal working via conventional gas engines, or the gas turbine which he proposed, would do the generating far better than the coal-boiler-turbine-alternator linear progression. A by-product, ammonium sulphate, would fertilize the land, and tens of thousands of men would be released from the filth and drudgery of 'carboniferous capitalism' in order to enter clean, high-productivity jobs. His eloquence took his audience to the very threshold of utopian optimism (for Britons, at least) when he said: 'we may acquire knowledge to control by electric means the sunshine that falls on our land'. In many ways this was Ferranti's finest hour, certainly it was the zenith of the early electric era. Now electricity, through one of its high priests, made absolute and far-reaching claims. Even though sceptics scoffed at much of what he said, Ferranti's claim was credible. Of course, managerial inertia, politics, natural gas, etc, can explain why Ferranti has not 'come off', but there was no fundamental mistake: electricity had made its great offer—it could sustain a modern industrial economy. The All-Electric age was nigh.

Electricity was its own propagandist in the Great War. It proved admirable for powering remote and strategically sited munitions factories, and the various Reconstruction committees were certain that the demand would continue into peacetime. Shortly after the war, however, in order to ensure that no hiatus did develop, various interested parties led by the British Electrical and Allied Manufacturers' Association (and especially Dunlop); the Incorporated Municipal Electrical Association; Institution of Electrical Engineers and the Cable Manufacturers Association, jointly set up the British Electrical Development Association, usually known by the acronym BEDA (1919). This organization symbolized the spirit of the later Electric Revolution perfectly. It boosted electricity by advertising, exhibitions, advice and demonstration.

Its style was adopted by the more aggressive suppliers, who used the 'hard sell' technique. High in this league was Hackney Corporation Electricity Department, which maintained vigorous propaganda through the 1920s, and waged a long war with the Gas Light & Coke Co. The Gas Co opened the battle by quoting the (wartime) National Kitchen Division of the Ministry of Food, which found 'electricity for cooking anything but favourable. Gas appliances are more constantly reliable and cheaper . . .' The rejoinder was quick: 'Hackney Electricity Department agrees with the Gas Company—an ounce of fact is worth a pound of claim . . . Hackney Electricity was never used in the National Kitchen',[7] the Department went on to quote letters of 'satisfied and delighted customers', including ladies no longer nonplussed by unpreparedness 'when the Minister calls . . . the almost tragic tenor of the housewife at these unexpected visitations entirely disappears when HACKNEY ELECTRICITY is used for lighting, cooking and cleaning in the home'.

Another cheerful and lively propagandist was 'SWS' (the Shropshire, Worcestershire and Staffordshire Electric Power Co), sure to be found at West Midland county shows, or adorning High Streets with blazing showroom windows. A singular effort of this company was at the Kidderminster Hospital Carnival of 1923. A 1-ton Edison battery lorry, mounting a tsar-sized iron advertised 'Electric Ironing at ½d per hour'. Two details were puzzling: the iron's handle mounted a small demon, complete with arrowhead tail. The driver was rigged out as the Archfiend himself, oddly sporting a court dress sword. All to do with *heat*, one supposes.

The Shropshire, Worcestershire and Staffordshire Electric Power Co was one of the leading rural electrifiers, thereby occasioning to a very large degree the social revolution by which rural England (and almost as much, Scotland and Wales) have become dependent extensions of urban England. The process was started by the railways, and finished by the national grid, and electrically-based mass media. The quality of small town life is still palpably different from that of suburban Birmingham, of course, but the margin is getting narrower. Within its designated area 'SWS' had connected three-quarters of all villages (of not more than 1,000 inhabitants) by 1930. Oil lamps and solid fuel ovens, which had served for generations, fell wholesale to the company's onslaught. It offered, *inter alia*, twelve quarterly instalments of two shillings (10p) per lighting point, or three shillings and fourpence (17p) per heating point. Cookers could be hired at five to ten shillings (25p–50p) per quarter; fires at five shillings (25p) *per annum*, but never kettles, which were apparently ill-treated. In this way the land of Housman and Elgar succumbed to the mobile demonstrators and blizzards of quite tasteful leaflets issuing from regional headquarters at Smethwick. One of these handouts showed proudly ('unashamedly' we would say today) a 1,300 ft double transmission line soaring over a valley, novel and unusual in 1930, and symbolizing very poignantly the dependence of old England on the enterprise and power of new, urban England: in this case the Black Country.

A great suburban pusher of electric power was 'Northmet', which served the Betjeman-land of middle class villas and light factories North of London. Here, in 1924, was erected the greatest and most transient monument to the Electric Revolution: the Wembley Exhibition of that year, correctly the *British Empire Exhibition*, designed to express the industrial expertise of the mother-country, and the strong solidarity of the Empire in general, although (few realizing it at the time) it was well into its late afternoon by then. The electricity industry, being new and science-based was given a prestigious place at Wembley, and it used this enthusiastically. The British Electrical Development Association was to the fore with an area marked 'Electricity in Service'. Describing this, the Director and Secretary of BEDA (J. W. Beauchamp) made some acute prophecies:

> The load charts of power stations [will be] future raw material for the Historians to whom they will explain many of the far-reaching changes now taking place in social conditions, and stand with fuel and food records in importance as an index of economic progress.

He spoke accurately, for power consumption *per capita* like income *per capita*, telephones, paper output, etc, has certainly become such an index. The exhibition contained an *embarras de richesses* of electrical equipment. There was an 'electric house' in tasteful California-Spanish design, serviced by a full array of gadgetry.

GEC had a giant 500kW rotary converter on view, Metrovick showed off everything from a dissected locomotive elevation, to electric fires. There was an asceptic and austere white 'electric bathroom', not far from the black incubus of an 'Armourclad

Reyrolle Switchgear'—all these artefacts showed how swiftly the industry had grown since 1880, and how, after a sluggish start, British electrical engineers were approaching omnicompetence by world standards. Even the hopeless medley of supply and distribution systems which lurked behind all this salesmanship and flamboyance had some artificial unity and coherence imposed upon it by the elegant, standardized signwriting whereby statutory undertakings had their names painted around the 'Faraday Gate'—they almost looked part of a system logically conceived and managed.

The electric house at Wembley was but one version of a largely suburban phenomenon that became popular and common *c* 1925–1935: the 'All-Electric House'.[8] Ferranti's All-Electric doctrine had one of its last major battles to fight on the domestic front. Some enemies were hard to dislodge; gas remained popular for cooking, with its very flexible heat control. On the other hand, the mass of suburban houses being erected at the time (and especially after 1931, in an era of low interest rates) were for small and servantless middle-class families. To the housewives in these areas, a vacuum cleaner or an electric fire was a most convenient agent of 'emancipation', and this was no catchword to young wives who remembered gas lighting, coal fires and elbow grease.

All-Electric houses were to be found in many places, usually they were standard new houses temporarily rented by the British Electrical Development Association or the local electricity undertaking, and open to viewing by the public. The Birmingham Electric Circle (of BEDA) took such a house in Court Oak Road, erecting a sign on the new arterial road nearby: 'To the Electric House, Admission free, open 2 pm-9 pm'. Birmingham Electricity Department informed visitors that they might have special domestic contracts whereby electricity would cost them 1d per kWh. From the generator's point of view, all this was a good way of 'load building', a phrase which was also common in technical circles at the time. The success of such campaigns was repeated in many places: Edinburgh, Cheltenham, Cambridge and Ipswich, to name a few. Dublin and Bristol had All-Electric Bungalows, and for consumers who preferred hard facts, there were Professor S. Parker Smith's papers which proved the superiority of electricity in the home, grounding the case in carefully collected empirical data rather than advertising talk.[9] Some of Parker Smith's evidence seems obvious or even quaint today: he found that the severe winters of 1928–9 reduced the effectiveness of his electric fire, until he blocked up the old chimney below which it sat. He pronounced electricity superior to coal and gas in every way, except that he got 'slight shocks from the kettle'. Still, his five-year controlled experiment confirmed what many were finding out—that electrical energy could make the 'logistic infrastructure' of domestic life cleaner and more convenient than ever before, and in a way that most people in the 1920s would have thought unbelievable when they were children.

There were also All-Electric restaurants and factories, popularized by BEDA and others; for example the 'Tricity' restaurant in the Strand, London (1927), which claimed that 'The All-Electric Restaurant is not a distant dream',[10] was fitted out as a showpiece by the British Transformer Co, and seated three hundred patrons. A typical All-Electric factory was Messrs Carreras cigarette works at Mornington Crescent, London. It went up in 1930 (in the Egyptian style then in favour) and housed over 400 hundred electric motors; it represented a fine load builder for the St Pancras borough electric power station, with a total demand of 1,000kW (see figure 24).

Among the groups propagating the electric gospel was the Electrical Association for Women (1924), backed by Ferranti. Towards the end of the Electric Revolution it presented some propaganda-by-drama: *Watts in the Home* (1930). The play appeared on the surface to be the chronological success story of electric power, told in a blow-by-blow account. Starting in 1880 the action revealed progressive changes in lighting by carbons, filaments and later gas-filled bulbs. The play also showed, on another plane, how the starry-eyed young things of the 1880s who welcomed 'the' electric light, had become reactionary fogies by 1930. *Watts in the Home* was not 'profound social comment' but it was highly significant in content and context. By 1930 electricity was established, and indeed, it had an Establishment as well. The All-Electric campaign had not achieved total success,[11] but by setting its sights high it had achieved a great deal at the expense of coal gas and steam. The need to propagate was still there, hence the play. But the boyish exuberance of *fin de siècle* and Ferranti in 1910, was passing away; indeed Ferranti himself died in 1930. The revolution was all but over, and electricity had arrived.

NOTES

1 See *Electrical Review* especially Vol 55.

2 ie Willesden: the incident referred to took place on the Grand Union Canal, and affected either 'Northmet' or Willesden Corporation.

3 See p 65.

4 The first periodical in electrical matters was the *Electrician* (1861–4), another of the same name started in 1878. The *Electrical Review* (originally the *Telegraphic Journal and Electrical Review*), started in 1872; the *Electrical Times* (originally *Lighting*) started in 1891, and was an effusion of Robert Hammond, the Brighton pioneer.

5 This embattled metaphor was significantly quite common, and favoured by the *Electrical Magazine*. It looked favourably on the battery-powered 'Electrobus' which would help electricity to 'assail' steam and petrol buses, 'if the campaign is generalled . . . it will carry the position by storm. The London Electrobus Co is the first division to take the field'—*Electrical Magazine*, Vol 5, No 5.

6 A ditty of the day, composed extempore by some Electrical engineers dining at the Midland Hotel, Manchester ran:

Ferranti, we all agree, has got a special plan
He's going to put cheap juice on tap for woman, child and man.

7 As if it were different! Cf AC being 'more lethal' for the New York State electric chair than DC.

8 All-Electric show houses were not an invention of the inter-war period. In the Manchester Exhibition of 1908 there was a 'Model House' (designed by Goodall, Lamb and Heighway) in whose kitchen 'the all potent . . . electricity is working wonders . . . unsullied by soot or smell, it is roasting, stewing, boiling, chopping . . . in fact allowing full swing to the genius of the cook'. In 1909 'Southmet' opened the All-Electric 'Electra House' at 111 Bromley Rd; this had a vacuum cleaner, knife cleaner, sewing machine, etc, all electrically powered.

9 See *Electrical Review*, 1929; *Journal* of the Institution of Electrical Engineers, 1924.

10 Quotation from *Electrical Review* of that year.

11 For example, in rural electrification. There were only 3,700 miles of rural power lines in 1929, but 30,000 miles by 1937; this *inter alia* powered 5,000 of Great Britain's 450,000 farms. As in other electricty supply matters, Great Britain's slow start enabled her to make spectacular strides later, with the newest methods available.

13, *Electrical Phenomena*

AS the Electric Revolution worked out its possibilities, so a stream of new phenomena appeared marking its progress. By studying this morphology of revolution much can be learned of the industry, and the society it served. A few examples will suffice; any attempt at all-inclusive comprehension would be portentous and staggering, interesting only to a handful of *afficionados.* Power stations, tramways and a few miscellanea will do the work admirably.

A walk in the earlier suburban or industrial districts of many towns can often reveal a few gems of 'early electrical' style. The maturer Edwardian streets, for example, still sport concrete or steel manhole covers bearing the name of the old electricity distributor in *alto rilievo.* Cast-iron boxes for feeders, isolators or transformers that once energized the local tramways can sometimes be spied, as can the iron or steel posts that carried the tramway 'overhead'—occasionally and beautifully adorned with wrought iron scrollwork, and what Hamilton Ellis called elsewhere '*pickelhaube* finials'.[1] Messrs Dick, Kerr of Kilmarnock, specialists in general tramway contracting, had a singularly excellent line in scrolled bracket arms, for which they charged extra; some of these posts have been retained, modified, for present-day highway lighting. By looking down the more lush pre-Great War roads, dripping with laburnum blossoms and shaded by taller trees, one can get a fleeting sensation of such roads in their full electric glory, when smooth steel rails set off the crown of the road, and arc lamps hung like white fruit, off-duty in the still summer afternoons.

To see a gang of men from the local electricity board, or a large contractor, laying cables is unexciting today, except perhaps to young children to whom it may be a fascinating novelty. Not so in the Electric Revolution (see figure 32) when new cables in porcelain conduits were uniquely mysterious and interesting. The neighbourhood often turned out to see them being laid in trenches. There was even a chance, later, that a faulty mains would misbehave after the event; one piece of evidence being powerful whiffs of bitumen coming up from between the pavement cracks. Another fiendish phenomenon was the water resistance, used by the Cleveland and Durham network. Their early 11kV power line ran over Eston Moor and had static problems; the water resistance effectively cured this, but was housed in remote substations where it gave off an eerie and vivid blue light, and great noise. Apparently the local youths gave these stations a very wide berth.[2] When electric lighting was in its first phase, another local phenomenon was the carbon man, who renewed the carbons of street lamps, which was done by lowering the lamp to ground level by means of a key in the column base. On quieter streets, incandescent bulbs were often used, and in these cases old gas lamp holders sufficed, being quickly converted with an extension which usually mounted a butterfly-shaped mosaic mirror.

Power stations were and are the most obvious specific buildings associated with the supply industry. Substations were deliberately rendered discreet,[3] especially in later years. The All-Electric suburbs hid their substations behind hardy bushes, or even fixed pseudo-Tudor slats about them to tone in with the similarly styled environs. Not so the power house! Power stations have been grand, poky, eclectic, gigantic and neat by turns, but never did they appear in the least degree *coy*. Whether they have been designed by architects or engineers they have been positive additions to the landscape, although curiously underestimated aesthetically, and rather taken for granted in view of their immense importance. Perhaps this is some way of sublimating our dependence upon them, which we are reluctant to admit. Suffice it to say they have been eschewed by the bulk of commentators, a fact which only contributes to their intrinsic mystique and fascination.

The earliest power houses (except for a few like Deptford) were rather cramped and even squalid. A major reason for this was the employment of low voltage DC which in turn necessitated building the station virtually amidst the property which it was to serve. This was usually where land values were high, hence the small station which was so common with early London electric lighting companies. A foreign observer, G. de Semenza distastefully reported of many British stations (1894):

> Let us visit one at random, no matter which for they are almost all of the same stamp. An unimposing entrance, a gloomy little back alley, and at the end the workshop of the station. A covered pasage, broad enough to allow a coal cart to pass, traverses the workshop. On the left hand side . . . is the coal store, on the right, up a few steps we find the boiler room. I call it boiler room for the sake of the word. There is a long row of Babcock and Wilcox boilers arranged along the passage . . . we enter the engine room . . . there are installed half-a-dozen or more Willans engines coupled to as many stout bi-polar dynamos . . . into the engine room a grey light falls from a skylight, the walls are grimy and between one machine and the next, or the wall, there is hardly room to pass.[4]

The need to extend power plant as demand grew usually meant squeezing another Willans set into an odd corner, adding to the congestion and muddle. The greatest single factor in breaking away from this pattern was the adoption of the compact steam turbine.

Within these early plants, whether powered by reciprocating or turbine sets, there was a provisional and boyish spirit that manifests the long lost *joie de vivre* of the Electric Revolution, of the type captured by 'Innocens' in his short stories. All this would be out of place in the clinical and serious affair of power generation today. In a classic and reminiscent description of those days

(C. A. Holbrow and Sir William Tritton in *The Engineer*, 1934) relating to the old Manchester Square (London) station of the Metropolitan ES Co, we hear of such incidents as:

> The second expansion blading (of the turbines) often suffered from what we then learned was the 'whipping' of the shaft. This did not worry us very much as it was a simple matter to cut out the damaged blading, close up the set, and continue working . . . sometimes one-third of the total blading would be cut out.
>
> It was a most extraordinary thing how long it was possible to keep one of the turbines going when a hot bearing occurred. The driver used to cover up the bearing with cotton waste, and we had a cold water circulation which kept the waste wet. Perhaps after an hour or more, the bearing was still heating, with a quantity of smoke coming out. We then used heavy cylinder oil . . . and this either had the effect of cooling the bearing down, or else the smoke became so great that we could not live in the engine room.
>
> The original alternators had surface-wound armatures with the conductors held down by piano-wire binding . . . frequently the piano-wire fractured and was forced out at the end of the machine. At such times there was a lot of wire whipping about . . . and a good deal of fire flying about. After one or two burn-outs the men in charge got to know by the variations of the hum of the machine when a burn-out was likely . . . and one man was then kept at the stop valve ready to shut down. No one else was allowed to be near.

Sir William Tritton, taking up the discussion, remembered the 'interminable trouble of balancing' the alternators, which would start: 'walking down' the engine room if unbalanced. It once took two years to rewind an armature and even then 'we never got the thing to run decently'. Switchboard troubles were legion; one led to a fire which damaged some generators, necessitating a change-over to some others—he concluded cheerfully: 'we did not bother too much about putting all the lights out in the West End several times a day in order to change over'.

At the maximum there were over five hundred stations in Great Britain, ranging from the vast to the shack-like. Their style and internal layout was very varied. The Stalybridge, Hyde,

Mossley and Dukinfield joint station[5] of 1904 was a very neat piece of architecture on the bank of the river Tame at Stalybridge, and its carefully shaped 225 ft chimney seemed sensible enough in the land of textile mills. Rotherham No 2 of 1919, on the other hand, squatted like a battleship on the banks of the black and oily river Don, with two short steel chimneys athwartships, and plenty of black corrugated iron clad about its frame. It was not beautiful, but delicately styled things would be out of place in such raw and spartan surroundings, amid the steady fallout of the local steel mills.

Inside power stations the clutter that Semenza noticed was cleared away after the Parsons era opened, although the high-speed reciprocating sets were often placed in neat serried ranks. The contrast between a reciprocating station and a turbine one was often dramatic; the old York station (1909) had its Willans sets in cheerless brick alcoves and was fairly typical of its type. On the other hand, the cathedral-like interior of Nechells (Birmingham Corporation's showpiece of 1926) allowed light to flood in through the roof, on to the well-groomed turbine sets singing in the aisle. Nechells was remarkable, also for the batteries of wood cooling towers that always seemed rather less obtrusive than the concrete ones which became popular in the 1930s.[6] Some of the earliest power stations cooled water by squirting it out of batteries of pipes, in great sprays: Neptune Bank at Wallsend, and Frindsbury, Rochester, used this primitive means; Frindsbury was in many ways the 'basic power shed', with a steel frame, corrugated iron walls; the whole set down among the cement works and barge yards of the lower Medway. There was also the Klein (later Clyne) cooler, made up of rectangular wooden box-towers with fan induced draughts. The huge fans spun round in the tower bases, more impressively alive or 'kinetic' than modern coolers; the two best-known wooden cooling towers were the Premier and the Visco. One London station, Bow, of the Charing Cross, West End and City Electricity Supply Co used a series of what looked like giant steel bottles; other variations were tried in the pre-Weir era, until the monumental concrete tower was adopted as standard for the larger power stations in vogue after 1926. Later cooling towers were more pleasantly shaped ('hyperbolic'), than the rather awkward high-waistline ones of pre-war years, towers which might mount enormous advertisements such as 'Use Electricity'[7] visible for miles from passing trains and cars. The concrete tower is more durable than the wooden one, and because of its greater size, fewer are needed. Yet, it was a memorable sight to see those wooden coolers, rank upon rank, steaming powerfully on an autumn day. Coming up Lisson Grove, Marylebone, one could see them around the Regent's Canal stations, like steaming and ethereal outworks, shrouding the quietly whining turbine rooms.

Because of the need for cooling water, power stations have tended to gravitate to river and canal banks. The London County Council station at Greenwich (1904) (see figure 29) on the South bank of the Thames paid deferential compliment to the architecture of the neighbourhood (mainly the superb collection of 'Naval Classic' buildings, from the old Observatory to the Royal Naval College three streets away, down river). Today, the very notion of placing a power station so close to these rare jewels of architecture would cause an outcry, but sensibilities were less developed in the Electric Revolution. As it is, the LCC station has blended into Greenwich scene, and is a time-mellowed old familiar on the Surrey shore. Up river is Bankside station, opposite the City of London, and most carefully designed (by Sir Giles Gilbert Scott, 1948) to enhance 'riparian London' rather than merely defer to it. Its most remarkable feature is the tall, square chimney stack, from which there ceaselessly bubbles pure-white fleecy exhaust.[8] None of the waterside power stations, however, can compare with the greatest 'Temple of Power'—Battersea, also designed by Scott (see figure 33). It was planned in 1927 by the London Power Co—a new amalgam of the type permitted by the

1925 Acts. After preliminary work, the foundation stone was laid on St George's Day 1931, partly as a centenary gesture to Faraday's aforementioned *annus mirabilis.*

Power stations offered, like railway stations, large-scale examples of functional architecture. They occasionally softened their contours with some carved brickwork, or terracotta, but generally have always been severely plain in outline. This stark and un-fussy style can be very attractive, especially in certain climatic conditions. Also, power stations were early users of the frame and curtain wall principle on the grand scale, like the Dublin Tramways power house at Ringsend (1898–1900) whose skeletal steel framework loomed up above the canal during construction, attracting much wondering attention. First of the steel-and-glass modern stations was Dunston,[9] built by NESCO to the designs of Merz and McLellan in 1931.

Although brick-clad Battersea looked rather traditional in some ways, it was very progressive in one. In order to mollify opposition to the building of so large a station in mid-London, the London Power Company had to agree to cleanse the station exhaust. As with Stonebyres (see p 101), Battersea was erected in a world that was showing the first signs of serious environment-consciousness on a large scale. The Electric Revolution sped on the arrival of the sort of society that would care about these things, by offering clean energy, and, partly, by helping to create the sort of society that had the time to raise its collective head from grinding toil to see the environment. Before allowing the Battersea scheme, the Electricity Commissioners insisted that it 'emitted clean smoke', and sulphur extractors were duly fitted to the exhaust system.

Chimneys are probably the most obvious external features of power stations, after cooling towers. They are many and varied to the interested eye, and have often become famous landmarks. Battersea, like Lots Road (and all stations with a chimney at each corner of a rectangle) suggests to unkind commentators an upturned animal. Fulham Power Station has four stacks *seriatim* like a Russian battleship *en route* for Tsu Shima, sailing over the plains of South London suburbia. The more functional approach led to rows of squat forced-draught chimneys, or pipes. Blackburn Meadows (Sheffield Corporation) had a battery of these austere stacks, but Ringsend (Dublin) was set off by two soaring steel chimneys, each 230 ft high, lap jointed and tapering upwards to flared mouths—altogether wonderful adornments to that bare, metallic station.

Inside stations, one met with varied layouts. Carville (1904) was the first station to have central control (see figure 27) for the nearby network,[10] truly the officer on duty kept, as the catch-phrase had it, 'the pulse of civilization going'. Suitably, many early power station turbine halls led the eye to a switchboard gallery that was a kind of high altar, approached up sweeping open staircases of iron. At the Wood Lane power station of the Central London Railway one approached this gallery up a central aisle, flanked by horizontal steam engines whose winnowing crankshafts drove alternators, set further back. It was as if the controls on the gallery at the end were guarded by machines *couchant*, ever ready to spring. Canterbury Corporation station housed some neat Brush-Ljungström sets; these were favoured by many small- and medium-sized stations, and were housed in a torpedo or sausage-like casing. This was in turn put on a plinth, over the condensers. So, the finished set seemed poised, again, as an altar—noisily whining above one's head like the *Lord of the Dynamos*, in H. G. Wells' excellent short story about macabre and grotesque happenings in an early power station: virtually the only foray of a secondary fuel industry into first-class literature.

Many of the local stations of the pre-Central Electricity Board era were pleasing buildings, like Port Dundas (Glasgow); Sutton (Southmet) and the first Brimsdown (Northmet)—all large by Edwardian standards, but middling by the 1930s. They were workmanlike and well proportioned, as were the much smaller 'local lighting' stations, like Chichester (see figure 31), very plain in red brick with a few white facings, and black-and-white chequer terrazzo flooring within—a great favourite at the time. Most pre-1914 stations were built in brick, usually for load-bearing but occasionally just for curtain walls. The colouring varied; the Isle of Thanet Electric Supply Co's station at St Peters was of a whitish brick, giving an eerie effect on misty evenings; Edinburgh Corporation's at Portobello was an arresting salmon red. Since nationalization stations have become rather plain again, in the old utilitarian tradition of Sheffield, Rotherham and Ringsend, the normal form of construction being a heavy steel frame with aluminium and fibreglass sheeting. Although they differ in style from Battersea, and are a reversion to the 'shed style' of Carville, etc, they embody one tradition which has been common to most stations for forty or more years: sheer scale. Power stations are a part of a broad-gauge industry that cannot afford to do things on the mean or penny-pinching scale, for there must be ample reserves of energy available. Modern stations all have the cathedral spaciousness of their Merzian predecessors, the same unremitting din of the turbines and alternators—the whole effect being what the Germans call *grossartig*: on the big scale in all dimensions, and to all the senses.

Whereas power stations have generally been underrated as architecture, they have never been so controversial as the grid tower or 'pylon'[11] which carries high-tension current about the kingdom. The earliest grids, in the North East, were at modest voltages, 11kV or 20kV, and so their towers (often creosoted poles) were unobtrusive dwarfs compared with the 400kV giants of later days, and the 'supergrid'. The controversy really started with the erection of the national grid, 1927–33. By 1931 the *Electrical Review* was sure that 'their benefits will outweigh their inelegance'. Although many found them not in the least inelegant, their sweeping progress through rural England sparked off a public row that became virtually permanent. To the environmentalists, then and later, came the same answer: underground lines are costly to lay and maintain. By 1970 the difference (per mile) was sixty-five thousand pounds (overhead) compared to nine-hundred and fifteen thousand pounds (underground) for 400kV; and ten thousand pounds compared to ninety thousand pounds for the old 132kV. No government wishes to be responible for the higher price of electricity, or the diversion of funds for general welfare purposes in order to bury the mains. In the 1930s the usual answer was to say that people would get used to the grid. So they might have done if the original 132kV system had frozen as at 1933, but it grew continuously in mileage and voltage. Really, the pylon controversy is the classic social paradox, like the hydro-electric fights of the 1920s and 1930s. People want a higher standard of living, but not the sacrifices which it entails. Occasionally the beneficiaries are distinguished from those who make a sacrifice (such as farmers), but generally the problem is one of a collective social schizophrenia. In any case, the 'ugliness' of the grid is not universally proven; there are times and places where it displays functional design at its best: swinging down into an industrial valley, with the late sun catching the wires and making them gold, or when towers loom out of a wet mist, like the Martian fighting machines of H. G. Wells. The discerning ear can be rewarded, too—by an interval of a third on still afternoons, or by the banshee-wail of 33kV lines in a strong wind.

A nice attempt to civilize the grid came from Arthur Mee, the journalist and editor of the *Children's Encyclopaedia.* When the West Kent Electricity Co wanted to put a mains under the drive of his house[12] in 1931, he gave consent, but instead of asking for the usual wayleave rental—mere lucre—Mee asked for 'a red rose, in June every year'. The West Kent Co dutifully complied, even sending the manager's little daughter to present it. This could be construed as quaintly precious, or it could have been years ahead of its time, the action of a man living beyond materialism. However, Mee's example remains unique, and is likely to remain so these thousand years.

While the pylons, and the electricity industry which they came to symbolize, have many years of secure future ahead, ironically, the electricity industry has sustained some other industries and phenomena that have led comparatively ephemeral lives: the electric trams, the trolleybuses, the widespread cinema-networks and circuits of the 1930s and '40s. The history of the electric tramway has been recorded in detail, and most admirably elsewhere.[13] Here they are considered purely as electric phenomena, for tramways were always very closely connected to early electric enterprise. Garcke's classic Manual for many years was of *tramway* and electrical undertakings; the *Electrical Review* devoted many pages to *Tramway Notes*. Next to public lighting, the electric tram was the most obvious and dramatic example of electrically-based progress *c* 1890–1910. The zenith year of the British Tramway was 1927, when there were 2,554 miles of route and 14,481 cars operating. The lion's share was in the municipal sector (1,862 miles; 11,969 cars), the rest was private, dominated by the British Electric Traction group. The total of municipal operators (not simultaneously) was 184, of which 112 actually ran tramcars, rather than merely owning track. The companies began to decline earlier, from 1920 (eight hundred miles; 2,850 cars) largely because they could not fall back on the rates in hard times; also they often served less promising territory, or ran isolated systems, unconnected with any meaningful networks. The first two tramway abandonments were well within the Electric Revolution, and both were satrapies of British Electric Traction: Sheerness (1917) and Taunton (1921).

The close connexion with the infant electricity industry arose from the fact that supplies needed to build up power station loads. Although it would be an exaggeration to say that electricity was ever 'dumped' on consumers while the tramways paid for the fixed costs, it is certainly correct to see the tramway power demand as a most useful weapon in the original war against coal and gas. Electric light was very expensive, even a luxury, until the 1920s, and was prevented from being more so partly by the tramway contribution to better load factors. Purely lighting companies had a worse load factor than tramway served areas, although the existence of industry also in those areas helped the pattern even more.

Thus British Electric Traction operated both the local lighting and trams in, eg Sheerness, Merthyr Tydfil, and Weston-super-Mare. The Isle of Thanet Electric Supply Co (see figure 28) ran both in the Margate-Ramsgate area, and the Midland Counties Electric Supply Co (part of the Balfour Beatty group) in the 'D. H. Lawrence country' of the Notts-Derby coalfield. In South Wales, the Llanelly & District Electric Lighting and Traction Co (another Balfour Beatty subsidiary) owned a tram-and-lighting undertaking.[14] The very name of the undertaking might reflect its dual nature, like the Musselburgh & District Electric Lighting and Traction Co, or the South Metropolitan Electric Tramways and Lighting Co (Southmet) whose power station at Sutton was described (1908) as supplying six hundred consumers in addition to the company's thirteen miles of tram tracks.[15] Municipalities did the same thing, for example the Rawtenstall Corporation Electricity and Tramway Works. It was quite usual, however, to run the two separately (using Corporation power, of course) relying upon the intricacies of council politics to supply the necessary co-ordination. The 'exception to prove the rule' was London, where the LCC ran a large tramway network and powered it from its central station at Greenwich, it was not, however, a supplier or distributor of public power in the normal way: only companies or metropolitan boroughs did that in London. In later years some systems grew so large that whole power stations had to attend to their wants, a matter which ought not to have arisen and could have been avoided if really large stations had been built in time. For example, Liverpool Corporation's ninety-mile system at one time (1914) consumed the entire power output of five 'destructor stations'. These burned public refuse, and were a fairly common type of municipal station before the Great War. The advent of super-power stations undermined their initial economic advantage. The idea of one central station was the best practice, proved by Merz and others. Anything which led to duplication and division of the load pattern was wasteful, and led to serious peak-hour economic problems.

Tramways suffered from many disabilities, such as having to maintain road surfaces, offering less flexible operating patterns than motor 'buses[16] and having to pay for the constant repair and upkeep of 'overhead', points, etc. In 1948 municipal electric power was nationalized, so corporations had no great incentive to maintain the electric tram or trolleybus *per se*. On the other hand, the tram and even more so, the electric trolleybus, set standards of smooth acceleration, pollution-free operation and crowd-shifting that have never been equalled: the trolleybus was all but completely silent, too. These qualities do not necessarily count for much in the market-dominated economy, however. The tram and trolleybus were liquidated on economic grounds. If external economies ever come to count for anything, then they may return.

Even by 1920 the tram began to seem *démodé*. A proposal to 'entram' much of the North part of County Durham in 1920 met with a variety of objections, including one that 'the day of the tramway is over'[17]—the Sheerness tramway had expired three years earlier.[18] The American 'Interurbans' lasted for an even shorter period, a fact which suggests that the lack of a co-ordinated tramway network in Great Britain was not a major factor in its decline. However, the matter is worth consideration, for it is another and important manifestation of the 'electric localism' which hit the supply industry so hard. The multiplication of operators, of varying degrees of friendliness, did not exactly make for a 'dynamic reticule' of services (see figure 3).

South Lancashire had forty-seven tramway authorities, but at least used a standard gauge (4 ft 8½ ins),[19] North Lancashire was built to 4 ft 0 in, however, so that no through running was possible at Bacup, where the systems met. Yorkshire displayed an unusual lack of commonsense (or a normal one of aggressive independence, if preferred) in this matter: Huddersfield had 4 ft 7¾ ins;[20] Halifax 3 ft 6 ins, Bradford 4 ft 0 in, Leeds, Wakefield and Dewsbury were all standard. Sometimes authorities squabbled over a detail and through services were stopped:

> Accrington: the tramways deadlock at Accrington and Rawtenstall is still in existence. It is estimated that each municipality is losing £300 per year in traffic receipts . . . Accrington, who supply the current and cars for the running of the Haslingden system, have been unable to come to terms with Rawtenstall for a through service from Accrington. The result is that passengers by the new electric cars . . . are compelled to change at Lockgate, the boundary of the two towns. *Electrical Review*, 1909.

Another piece of fatuity was at the *Rising Sun* Inn where the Leyton and Walthamstow systems met—eight inches out of alignment, to prevent through running. On Teesside, the Imperial Tramways Co ran an integrated system, when this was 'municipalized' Stockton and Thornaby insisted on running their system separately from Middlesbrough, to show their independence. Through running continued, but a perfectly sound integrated system was cut up gratuitously.

What had started as a very justifiable cause of municipal *amour propre*, a new and fine amenity moving down the 'ringing grooves of change',[21] often ended up in the 1940s (if it survived that long) as rather shabby and a butt of rude jokes and acerbic comments. However, back in the Electric Revolution, the electric tramway was one of the most progressive products of the age, the pride of many a provincial culture. Trams should always, and rightly, be associated with the Northern 'industryscape', with the moans from their worn, flat-faced reduction gears echoing over the wet cobbles of Bacup, Osset or Rawtenstall; aeons before the 'development areas' or the 'affluent South' had become accepted parts of the British way of life.

NOTES

1 Ellis used this phrase in connexion with Midland Railway signal posts and boxes. The cost difference between plain and fancy tramway posts can be seen in estimates Dick, Kerr sent to the Hill of Howth Tramway, in Ireland: (for 294 poles) £6,937 the lot, plus £4 15s per post extra if decorated. The decoration was declined, see R. C. Flewitt: *The Hill of Howth Tramway.*

2 See *From Telegraphs to Nuclear Power*, published by Teesside sub-centre of IEE.

3 When Lord Moyne had bulk supply laid on to his establishment at Bailiffscourt, near Climping, Sussex, the Bognor Gas and Electricity Co moved a complete sixteenth-century cottage some fifteen miles for conversion and use as a substation, so as not to spoil the area.

4 In *L'Elettricista* November 1894; also in the *Electrician*, 1894.

5 See figure 35.

6 When Nechells was being demolished in 1970, these towers caught fire and blazed uncontrollably for a memorable night: they were completely lost.

7 Northmet was fond of this; Southmet emblazoned its early Sutton station with elephant-bold white characters 'Electric Power House', rather stating the obvious.

8 The chimney contains four flues. Bankside was planned to fit in with long-term development of the area. Battersea, in addition to its pollution-free exhaust, supplies nearby flats with hot water, in the Pimlico District Heating Scheme—Scott's two masterpieces enhance their surroundings in many ways, and set the highest standards of environmental planning.

9 The second station on the site.

10 See p 59.

11 See Appendix I.

12 Eynsford Hill, which looks down the valley of the Darent.

13 See especially Bett & Gilham: *Great British Tramway Networks*, and Klapper: *The Golden Age of Tramways*—the latter has a useful bibliography; also E. Jackson-Stevens: *British Electric Tramways.*

14 The nationalization of the Llanelli undertaking in 1948 led to the peculiar situation whereby the South Wales Electricity Board owned a trolleybus system (as successor to the trams) which remained the case until abandonment in 1952; a similar fate befell the Notts-Derby system, which however, was passed on to the British Transport Commission and the state-owned Tilling Group.

15 *ER* 1908.

16 This point is rather over-rated, since major traffic flows are fairly fixed in nature. On the other hand, one broken-down tram can cause a mighty hold-up in the rush hour!

17 See G. S. Hearse, *The Tramways of Gateshead*; *The Tramways of Jarrow and South Shields.*

18 The last complete system to be built was the Dearne District Light Railways (1924), a failure that was abandoned after nine years.

19 The longest possible journeys on this network were Liverpool–Bacup (59 miles), and Seaforth Sands-Hazel Grove, Stockport, (51½ miles). Charles Klapper *op. cit.* estimates eight hours (and plenty of changes) for the latter, including waiting time. Only for tram-fanatics, and especially so in wintertime!

20 So arranged in order to allow railway waggons to use grooved rails, Portsmouth and Glasgow did the same, although no waggons did so use Hudersfield's system.

21 Tennyson, *Locksley Hall*, apparently the poet was short-sighted and imagined that railway rails were *grooves*, like tram tracks.

14, Lost Innocence

Is it a fact—or have I dreamt it—that by means of Electricity the world of matter has become a great nerve, vibrating thousands of miles in a breathless point of time?

Nathaniel Hawthorne, 1849.

ELECTRICITY was the basis of much optimistic exaltation in the early years of the Electric Revolution, although more sceptical observers, like Henry Adams, while accepting the vast power of the new energy, were not led by selective amnesia to overlook the darker and turbid side of history, and the possibility that electric power could serve evil ends.

The electrical propagandists played such matters down, usually in all sincerity, and assumed that electricity was a power for good. Sharpened in conflict with vested interests, their crusade was marked with great moral fervour. Ferranti, Crompton, the British Electrical Development Association and the publicity-conscious power companies like the Shropshire, Worcester and Staffordshire Electric Power Co, were all confident that they were retailing something good to the community. Their new energy changed the world at a pace, and in a way that still seemed exciting in 1930. Yet in a sense, nothing changed. Human nature remained a constant. In another sense, everything changed, because culture, commerce and the political environment had their basis metamorphosed by electric power.

To many positivists and socialists (and not a few apolitical engineers also) the greatest burden on the new power was its hopeless distribution pattern. Once that could be rationalized, society could enjoy a new potential. This was not to be, however: in learning this the more open-minded rationalists matured politically. The nationalization of the whole industry in 1948 did not remove it as a political issue, rather it put it in the arena more clearly and firmly than before. Twice the industry was reorganized, and always it was pressed about by some kind of fuel policy. All the time intervention and *dirigisme* intensified; the Central Electricity Generating Board was ordered for many years to burn coal, and so forced, with increasing reluctance, to be a main prop for the declining coal industry. In its 1969 report, the CEGB observed (with an eye on the then Ministry of Technology): 'major plant decisions . . . must be aimed at securing the cheapest generating costs'. The fuel policy White Paper of 1967 acknowledged the problem, and announced the run-down of coal, but the fiscal burden on oil remained, in order to make it artificially unattractive to customers, including the CEGB. The state machine, operated by politicians who are sensitive to public opinion has very wide responsibilities, and must take a broader view of the power problem than the committed engineers. In 1970, Lloyd George's dictum about 'politics, not engineering' still has great relevance.

The future society anticipated by the electricity-men of 1880–1930 differed according to the character of the prophet. It is significant that so many of these engineers did speculate about the future, for that is the manner of revolutionaries who imagine that they are 'on the side of the future'. Tesla's social philosophy was austere and mechanistic. He saw man as a mere 'meat machine' determined by forces at present beyond his control. Future society might well overcome this, and man would be the Lord of Creation; his mastery would come from the use of electric power (see p 178). On the other hand, man would live for work alone, like the bee-civilization which he so much admired: 'vast desexualized armies of workers whose sole aim in life is hard work'.

This work-oriented society differs from most other electric utopias that have been suggested; we are reminded by Buckminster Fuller that improved technology enables us to do 'more with less'. This is also what Ferranti saw in 1910, when he posited that all primary fuels went into the making of one secondary fuel only: electricity. This would make society and industry cleaner, lessen the need for the more brutalizing types of toil and conserve valuable resources.[1] Conventional machinery could do the job, but like other contemporaries he searched for a workable gas turbine which, he was convinced, would be the talisman to open the electric future.

Crompton went in the opposite direction to those who imagined a future in terms of centralized grids, or swarming hives. As benefited the stalwart DC individualist, he foresaw an electrically-based cottage industry system 'evenly spread over the kingdom', and therefore a decentralized society. Lest Crompton be dismissed as a bucolic reactionary, it should be remembered first, that Kelvin and Tesla (in his later years) were also DC men, at least in terms of a large-scale DC network; and secondly that such networks, whether AC or DC do enable industry to be decentralized if it is thought economically, socially or strategically desirable. The galaxy of satellite towns, light industrial estates and the whole 'regional policy' of recent years rest on this fact.

Civilization is now more completely energy-based than at any previous time. Many futurists sensed this state of affairs coming seventy or so years ago. Of all available forms of energy it would seem to have the most assured future. This was noted in the high years of the Electric Revolution by R. H. Deane, writing in the *Electrical Magazine* in 1909. A fitting ending for this book is Deane's cosmic prophecy, breathing the spirit of those years, and outlining a possible future to which we are moving. Deane's voice is the authentic one of an electric revolutionary, encompassing future ages with his vision: '[electricity may] by the use of tidal energy ceasing only with the motion of our universe—

and hence truly corresponding with the "energy of the World"—be our salvation from the dark and cold that must otherwise settle on us and exterminate us'. Deane's cosmology may not be impeccable to fastidious minds, but one thing is quite clear: here spoke a man who knew that the future of civilization belonged to electricity, and he was speaking only thirty years after the Electric Revolution began.

NOTE

1 An interesting variant on the electrically powered future that came nearer than most to fruition was the 'World Corporation' of King C. Gillette, inventor of the safety razor. He planned a huge hydro-electrically powered city near Niagara, composed of vast barrack-blocks of flats, central kitchens, rationally planned industries and services and 'labour battalions'. He registered a company for the purpose and unsuccesfully attempted to get Theodore Roosevelt as company president. KCG put this and similar ideas forward *c* 1895–1910, see W. H. G. Armytage, *Yesterday's Tomorrows*, p 246.

Appendices

APPENDIX I: The Electrical Vocabulary

A certain sign of the relative newness of a social experience is in the initially haphazard nomenclature surrounding it. The quicker that the experience is accepted as commonplace, the quicker usage will settle down. This in itself shows how rapidly electricity 'arrived'.

Electricity comes from the Greek *electrum*, or 'amber', so called because amber, when rubbed, exhibits magnetic properties. Originally 'electricity' meant that which attracts other bodies, and was so used by Sir T. Browne (1646), he also invented electric (k) as the adjective, and as an alternative noun—we still use it as such: 'he is good at electrics'. In the modern sense, ie some independent force lying in things it was first used by Newton (1675), so by 1747 John Wesley could say that he had seen some 'electrical experiments' in a sense that we would use, or almost so. The difference is that the modern word tends to refer to a 'flow' of electricity, and in this sense it was first used by Benjamin Franklin, to whom we are also indebted for *electrify* (1747) and *electrician* (1751)—*electrification* dates from about that time also (*c* 1748). The adverb *electrically* was produced by Davy in 1812, by which time 'electric' was coming to mean, colloquially, 'sudden' or 'instantaneous'; 'the electric flash from the melting eye' (Coleridge). In the nineteenth century the word and its variations worked out their possibilities: electric light (1843); electric current (1879) and electric clock (1884). *Electromotive*, which sounds very modern, is in fact another invention of Davy (1806). Some variations fell by the wayside during the electric revolution, *eg electrize* and *electricize.*

Thus, Charles E. Lee in *The Metropolitan District Railway* notes that when that line was going to be electrified (1901) even in that one year it was going to undergo both 'electrolization' and 'electrization'. By 1905, however, 'electrification' was generally accepted. The early mounting for an electric light cluster was an *electrolier*, (Cf chandelier-candles; gasolier-gas jets) dating from 1884. The term gradually died away, except for use by perfectionists, and is used, if at all, to describe equipment made contemporaneously with the age in which the word was used.

Juice is an Americanism (*c* 1890), as is *Sparks* (for an electrician)—it was used to refer to wireless telegraphists in particular *c* 1915. Juice, of course, reflects the Franklin and now general idea of electricity as a flow. *Power Stations* (1901) were *Power Houses* (1890—and probably earlier) at first, but the 'house' is an obsolete usage, rendered so by the immense size of later stations. Engineers usually talk of 'grid towers' or transmission towers, whereas the public has always called them *pylons*. Strictly, a pylon is a solid stone pyramid-tower, sharply tapering. Obviously the grid towers look like that in outline, and the term is here to stay.

Traction, in the sense of powered transport or vehicles, dates from 1822, but its singular association with electrically-powered trams or locomotives is another Americanism, *c* 1885.

An interesting early usage, soon dropped, was to refer to *the* electric light, employing the definite article. Thus hotels boasted in advertisements that they were 'lighted by the electric light throughout' until the early twentieth century. This usage was obsolescent by 1900.

So whereas electricity is a word of great antiquity, and was inherited by scientists, engineers and suppliers, the alternative variations had all sorted themselves out by 1910—say thirty years after the revolution started.

Measurement

The need to standardize some system of units by which electricity and its effects might be measured antedates the Electric Revolution, and belongs to the 'age of preconditions', to adapt W. W. Rostow's famous terminology. Dunsheath[1] notes that the 'system' was rather arbitrary until *c* 1850, but in 1851 Weber suggested some fixed standards of electric and magnetic measurement. Ten years later the British Association set up a committee (containing among others William Thompson, later Lord Kelvin, and Wheatstone of 'bridge' fame) which reported up to 1870, and established the *ohm* as a standard of resistance, and the continental units of measurement in research and experiment. These had already been legalized as optional in 1864, over a century before the hoo-ha about 'going metric'. From 1891 *ampères* (amps) and *volts* replaced Webers and 'equivalent Daniell cells'; in 1894 the Board of Trade legalized the internationally-agreed values of these standard measurements. All these words were accepted as scientific terms a long time before they were adopted for general industrial and social purposes. Scientists had done all the necessary groundwork for over seventy years before the Electric Revolution, but by 1930 many householders and not a few housewives were conversant with the amp and the watt.

NOTE

1 Percy Dunsheath : *A History of Electrical Engineering.*

APPENDIX II : Forbes, Ferranti and the First Power Cut

Power cuts, whether from faulty machinery, labour stoppage or fuel shortage have become unusual in the years after the Electric Revolution, although the confusion when they occur (1947, 1970) is considerable and memorable.

Periodic fears are expressed, and occasionally borne out, that each successive winter will bring excessive demands for power, and outstrip the capacity of generating plant. The matter is taken very seriously in governmental circles, and a special committee of the Cabinet has been formed on occasion to investigate, and if possible avert, the crisis.

In the early days of electricity, power cuts were more frequent, and although just as inconvenient as later ones, they were not so widely catastrophic in an economy that did not depend all that heavily on electricity. The first major power cut occurred in November, 1891. A long period of fog and frost in London put a great strain on all distributors; none struggled so courageously as the London Electric Supply Corporation, and its central station at Deptford, to answer the continuous and high demand. The equipment could not stand the pace, however, and four days' total breakdown followed. Custom was lost, consumers and shareholders were furious. The London Electric Supply Corporation's chairman, James Staats Forbes (see figure 23), confessed to the shareholders: 'The whole thing came to a collapse, the dynamos, mains and everything went wrong, and for four days or more we were without light.'

Forbes was a past master at the art of soothing shareholders, and had great experience in the matter with his other companies, the London Chatham & Dover Railway, and two electrical pioneers: the National Telephone Co and 'Ediswan'. Ferranti, the chief electrical engineer of the London Electric Supply Corporation had left before this November chaos, after a succession of disagreements with the board. In their last row (after which Ferranti went) Forbes remarked: 'You are a very clever man, Mr Ferranti, but I'm thinking ye're sadly lacking in prevision.' This is often quoted against Forbes (with the benefit of hindsight, of course) because Ferranti was in many ways years ahead of his time.

But Forbes was no fool; he was an industrial diplomat of a high order, a believer in electricity and its innovations, and an enterprising backer of promising new ideas that others eschewed. As chairman of the LC&DR he helped to introduce a superior form of interlocking signalling system, which was safer than most of its contemporaries, and he supported the unusual twin-hulled ship *Calais-Douvres* which went a long way to conquering seasickness on the Channel crossing. Also, it was Forbes who gave Ferranti his chance at Deptford, and backed him for years when more cautious men would not have done so.

APPENDIX III : Wind Power

Although Great Britain lacks the necessary land-and-climate relationship which would enable large-scale hydro-electric power to be developed, it has often been suggested, on the other hand, that windmill-generated[1] power might be a useful substitute. A single, but very considerable snag presents itself: wind power cannot be stored, as water power can in a reservoir. Even the hydro-schemes that worked by the fall of a river, like Chester or the Falls of Clyde, at least had a steadily flowing river to count on, and to tap. Wind is more fickle, and short of employing prohibitively expensive batteries (and so tying the system to DC) there is no easy solution.

During and after the Electric Revolution, the matter was carefully explored, particularly in the flat and windy regions of the Baltic littoral. Britain, too, has investigated and exploited wind power on occasion. J. G. Childs & Co of Willesden Green offered a fairly comprehensive wind power service in 1909. The rosette sails of its mill (up to 100 ft diameter) were mounted on a 50 ft lattice steel tower, and guided into or out of the wind as desired by three related tails. These tails automatically veered the machine out of the wind by the amount requisite with the generation of a given amount of power. The 130V, 2kW generator worked from 800–1,600 rpm, and lay at the bottom of the tower. The whole apparatus cost one hundred and fifty pounds, and would give about 1,500 units of electricity per year. It was mainly for farmers in remote parts, so differing locations had very different energy potentials. A 6 mph wind would set the 'wind turbine' off, but constancy, not velocity, of wind was the great problem. At the Royal Agricultural Society Show in 1909, Childs lit their own stand and the next door one (Country Gentlemen's Association) with a 24 ft wind turbine, charging a twenty-eight-cell Pritchetts & Gold Battery. It could operate 'electric cooking apparatus . . . fans, radiators, flat irons, an electric lift, butter churns, cream separator, deep well pump, circular saw, incubator, etc', reported the *Electrical Review,* adding drily: 'It is to be hoped that Messrs Childs will be favoured with an adequate supply of breeze.'

During the post-Second War fuel crisis, attention turned to wind power again. With the existence of the national grid an accomplished fact, it was planned to generate the power at favourable locations, and transmit it to areas of high demand. The wind station chosen was Costa Head, in the Orkneys, and an experimental 100kW machine (designed by John Brown & Co) was duly erected there. The (then) Ministry of Power supported the scheme which came under the aegis of the Electrical Research Association. The chairman, T. G. N. Haldane observed[2] 'engineering problems proved more difficult than expected, the generators having to stand up to extreme gale conditions on very exposed sites', larger machines might have done the trick, but the fuel crisis passed, and in due course nuclear power was backed as the 'energy of the future'.

Wind power may still have occasional promise in remote places, especially in developing economies. However, in Great Britain it was another of the interesting ephemera of the Electric Revolution, and it enjoyed only limited application.

NOTES

1 The classic type of windmill was used on at least one occasion as an electrical generator. This occurred at Harleston, Norfolk (1907–12) when a mill supplied three houses with 50V DC; the mill was specially built for electricity generation, but was based on the traditional Norfolk windmill designs. Its career ended after a severe gale, and an oil engine took over. In a similar way, an old watermill served to generate electric power at Holford Glen in the Quantock Hills, Somerset, and Chagford, Devon, was lit by a water-wheel generator in 1891.

2 In correspondence with the author. Mr Haldane was chairman of the ERA committee on wind power.

APPENDIX IV: The Cinema

From its earliest days, the cinema was an interesting and significant example of an electricity based industry, and in that it was the first electrically based mass entertainment its story is 'pregnant with meaning' for the future. Its locations, artistic methods and its very existence were generally electrically-influenced.

The location of studio film production in Great Britain has largely been confined to London and the Home Counties. This was not merely because electricity was more readily available there—such a determinant would have allowed film studios to be set up in any major centre of population by 1900. London had many other attractions to the producer as a source of capital, actors, film and apparatus manufacture, potential large audiences, and increasingly the fact of being Britain's major cultural centre. Given these determinants, adequate electricity (especially for lighting) enabled the industry to settle near the metropolis with little trouble. A list of some important or early studios will demonstrate this:

Studio	*Ownership*
Boreham Wood & Elstree	National, Neptune, ABC, M-G-M, Burlington
Clapham	Phonofilm
Cricklewood	Stoll
Croydon	Cricks & Martin
Denham	London Films, later Rank Organisation
Ealing	Barker, Ealing Studios
Fulham	Cricks & Martin
Islington	Paramount, later Gainsborough[1]
Isleworth	G. B. Samuelson
New Southgate	R. W. Paul, 1899—a strong claim to be the first sizeable one: Paul was an electrical engineer and scientific instrument maker of distinction
Pinewood	Rank
Selhurst	Clarendon
Shepherd's Bush	Gaumont
Surbiton	Stoll
Twickenham	London Films
Walthamstow	Broadwest
Walton-on-Thames	Hepworth
Welwyn	ABC

There were also small studios at Beaconsfield (Wallace); Catford (Windsor); Kew Bridge (Prince's); and Whetstone (Zenith). Although many films were made outside this London area (eg F. S. Mottershaw's studios at Sheffield) there is little doubt that the capital was overwhelmingly dominant in this field.

Sometimes studios generated their own power. An early producer, Cecil M. Hepworth, searched the Thames Valley for a villa-in-grounds with power laid on. He ended up at Walton-on-Thames with the ideal house, but still no power. He installed a gas-engined generating set that supplied his studio's needs until public supplies were connected. On a much larger scale, Sir Alexander Korda's masterpiece—Denham Studios (1936) were served by Britain's (then) largest private power plant, with an output of 4,500kW, which could theoretically supply all stages, sets and ancillary services at once, although it was apparently never called upon to do so. Denham was proud to be 'independent of the grid', and clearly this was seen as something of a status symbol even though the grid was but three years old at the time.

The earliest generic name for the cinema was 'Electric Theatre' (USA 1902), a term which implied modernity and respectability. The first film shows were often fairground spectacles, powered by generators mounted on traction engines. By 1910, however, the electric theatre (see figure 37) was firmly established on permanent sites, such as converted shops or chapels[2] and later in specially built cinemas. The first 'super cinemas' came in the mid-1920s and were distinguished by great size[3] and occasionally arresting architecture. Cinemas were nearly always electrically lit and powered, although some early ones used lime and gas-light; secondary gas lighting circuits for emergencies were not uncommon, and were later obligatory. Electricity undertakings encouraged this new 'load spreader', for example by favourable tariffs such as the Hastings one quoted on p 40. By the 1920s there were some four thousand cinemas, representing a strongly entrenched element of mass entertainment—on a scale never attained by the live theatre. Cinema owners gave excellent value for their entrance charge, and assiduously built up a mass market by self-imposed censorship of material likely to offend audiences or articulate pressure groups, and by advertising copy which emphasized respectability, such as that of the Electric Theatre at

Peckham which ran shows to 'elevate, instruct and amuse' the 'Ladies and Children specially invited'. Admission charges were (pre-1914) usually threepence and sixpence (1p, 2½p), although the Electric Palace, Marble Arch, charged sixpence and one shilling (2½p, 5p), perhaps because it was the 'Picture rendezvous of the *haut ton*'. Whatever local exceptions there may have been, the cinema catered largely for the working class audience: the Hulton Readership Surveys (1950–5) found that 80 per cent of cinema goers were in that category.

A cinema could generate its own power by gas-engined sets, although most used the local public supply. In later years their patronage and numbers have fallen from the post-Second War all-time highest totals:

Year	*Cinemas*	*Patrons per year*
1946	4,709	1,635 millions
1950	4,660	1,396 millions
1960	3,034	515 millions
1968	1,631	237 millions

However, this shrinkage does not alter important established facts: that the cinema was the first true mass entertainment industry, and it was electrically based. Its greatest rival, television, is also electrically powered, and not a few latter-day entertainments have an electrical base also: pop concerts, record manufacture and playing, evening football, etc.

Broadly speaking, cinemas could be erected anywhere that electricity could be generated or supplied—literally anywhere. In this way an astronomical mileage of film helped to bind Britain into a more coherent and urban-centred culture. Since many films emanated from Hollywood, the sort of values and fashions enshrined in Hollywood films were known (and were perhaps influential) in Caithness, Cornwall and Wolverhampton. In this important way also, electricity welded 'one Britain'.

In a profound manner, electricity has helped to revolutionize culture and a whole way of life. The chronological coincidences occur again, and thereby strengthen the case for the concept of an 'Electric Revolution'. By 1909 the state felt that it was necessary to regulate the exhibition of films (from the point of view of safety, not ethical content) and did so by the Cinematograph Act, a contemporary of the Electricity Supply Act of that year. The industry formally censored itself from 1912 (British Board of Film Censors) and the term 'cinema' came into general use from 1910, another example of pre-1914 stablization of nomenclature of the type described in Appendix I. By 1930, the end of the Electric Revolution, the 'talkies' had arrived. Although the cinema had new aspects to unfold thereafter, 1930 marks, through the cinema, the arrival of modern mass entertainment, an important fact in the lives of all those who inhabit industrial economies.

NOTES

1 Ironically in an old generating station.
2 Thereby inheriting the organ so beloved of cinema owners and managers. In later years electronic organs were installed ('mighty Wurlitzers', etc)—electric music in electric theatres.
3 Britain had Europe's largest cinema; the Gaumont State, Kilburn.

APPENDIX V: Pre-Weir Scotland: Two Case Studies

Two regional examples will demonstrate the statutory and technical complexity of electricity supply that held sway at the time of Lord Weir's Report in 1925; and indeed continued until nationalization in 1948. The two demonstrated here are South West Scotland, and Fife.

South West Scotland

The dominant undertaking in these parts was one of the original power companies, dating from 1901: the *Clyde Valley Electrical Power Co.* It started operations in 1905, and its choice of a low frequency (25 Hz) determined much of the supply of the region, until standardization under the Central Electricity Board in the early 1930s. However, some undertakings went their own way with 50 Hz, and so, after years of being the local exceptions, they were statutorily transformed into precursors of the new era! The Clyde Valley Co owned three power stations: Yoker (1905); Motherwell (1906) and Clyde's Mill (1916); it also controlled the *Strathclyde Electricity Supply Co* which held a number of valuable Orders for lighting and power including *inter alia*, Clydebank (1901); Kilpatrick and Milngavie & District (1906 and 1922); Rutherglen (1905); Renfrew and District (1922) and Renfrewshire Western (1925). Similarly, the Clyde Valley Co supplied bulk power to other undertakings, which then transformed or converted it for distribution: *Motherwell and Wishaw*, a combined Burgh dating from 1920, before that Motherwell obtained its Electricity Order in 1895, Wishaw in 1904. Their separate histories were electrically reflected in 1925: Motherwell used 460V and 230V (DC); Wishaw 480V and 240V (DC); Motherwell had its own generating station, Wishaw was supplied with Clyde Valley bulk power.

Lanarkshire County Council was unusual in Great Britain, a county electricity authority (Dumfries-shire and Dunbartonshire were others) and it held a number of Orders from 1903. It had a small 240kW destructor plant, otherwise power came from the Clyde Valley Co. *Coatbridge and Airdrie Electricity Supply Co:* formed in 1906 to take over the undertaking of the Scottish House-to-House Electricity Co of 1890 (see p 73), and in 1925 it was an associate of the *County of London Electric Supply Co*

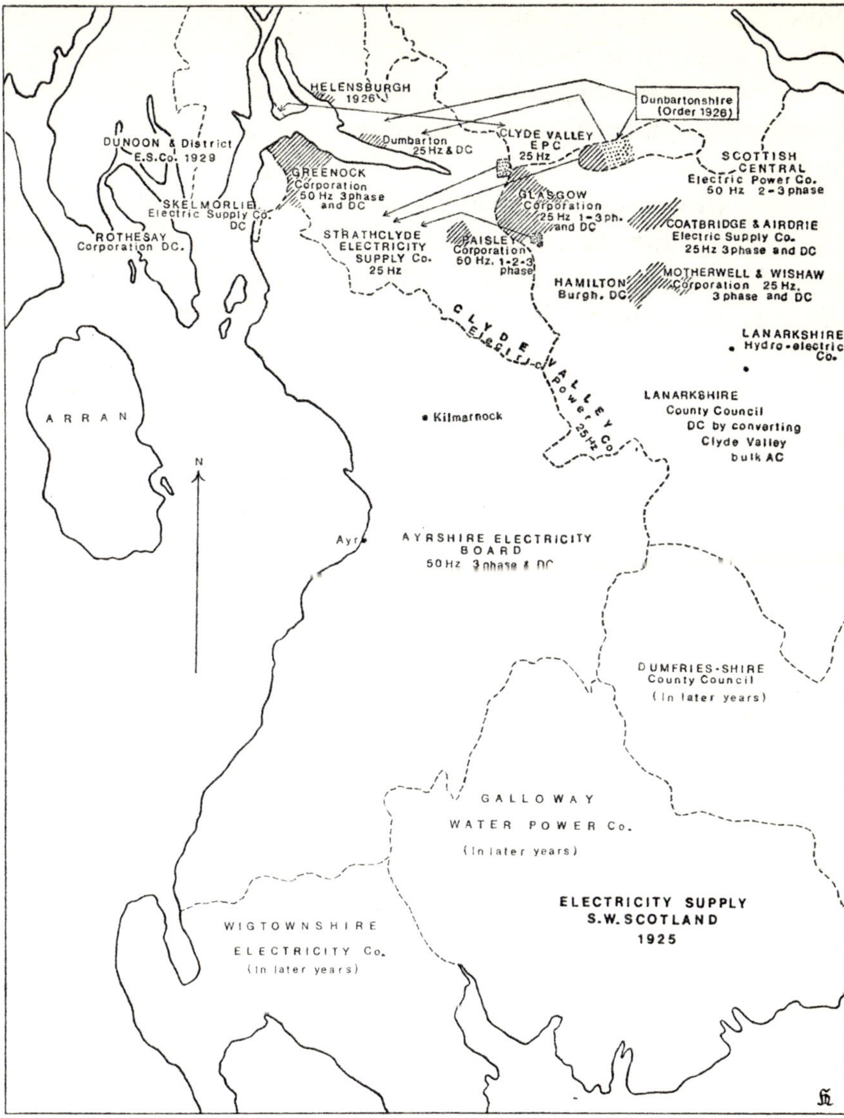
HELENSBURGH 1926
Dunbartonshire (Order 1926)
DUNOON & District E.S.Co. 1929
Dumbarton 25Hz & DC
CLYDE VALLEY EPC 25 Hz
SCOTTISH CENTRAL Electric Power Co. 50 Hz 2-3 phase
GREENOCK Corporation 50 Hz 3phase and DC
SKELMORLIE Electric Supply Co. DC
ROTHESAY Corporation DC.
GLASGOW Corporation 25 Hz 1-3ph. and DC
STRATHCLYDE ELECTRICITY SUPPLY Co. 25 Hz
COATBRIDGE & AIRDRIE Electric Supply Co. 25Hz 3phase and DC
PAISLEY Corporation 50 Hz. 1-2-3 phase
HAMILTON Burgh. DC
MOTHERWELL & WISHAW Corporation 25 Hz. 3 phase and DC
CLYDE VALLEY Electric Power Co. 25Hz
LANARKSHIRE Hydro-electric Co.
ARRAN
Kilmarnock
LANARKSHIRE County Council DC by converting Clyde Valley bulk AC
N
Ayr
AYRSHIRE ELECTRICITY BOARD 50Hz 3phase & DC
DUMFRIES-SHIRE County Council (In later years)
GALLOWAY WATER POWER Co. (In later years)
WIGTOWNSHIRE ELECTRICITY Co. (In later years)
ELECTRICITY SUPPLY S.W. SCOTLAND 1925

which held a mortage on it. Further evidence of the complex mysteries of organization and finance that characterized British electrical enterprise in those days.

The *Electricity Supply Corporation* was part of the Crompton 'empire' and ran eleven small undertakings, in this case for *Dumbarton Corporation.* On the other hand, *Dunbartonshire County Council* which was to get its Order in 1926, bought bulk from the Clyde Valley Co. *Hamilton* got its Order in 1898, and was set going by another 'empire' *Edmundson's Electricity Corporation*, which, however, later transferred the undertaking back to the Burgh. This was a DC system, with its own power station, although there were arrangements to take 500kW from the Clyde Valley Co.

Paisley bought some bulk power from the Clyde Valley Co, and had an Order dating from 1891, although supply only started in 1898. However, Paisley had its own power station, (6.3MW) and represented the Eastern outpost of the 50 Hz systems, the others being *Greenock* (Order 1883) and the *Ayrshire* joint board (q.v. below). Greenock also supplied Port Glasgow (Order 1913) and Gourock (1923). The *Ayrshire Electricity Board* was a joint undertaking, consisting mainly of the Burghs of Ayr and Kilmarnock, with a representative from Ayrshire County Council. It dated from 1924, in which year Kilmarnock shut its tramway and started to send power to Ayr by 20kV overhead lines. In turn, Ayr shut its power station (it kept its tramways until 1931, however) and thus this small system effected some very neat pre-Weir rationalization. There were interconnections with the other 50 Hz systems at Paisley and Greenock.

Glasgow Corporation was a 25 Hz system, with power stations at Port Dundas, Dalmarnock and St Andrew's Cross. It supplied plenty of 250V and 500V DC, much of it for Glasgow's famous 135-mile tramway network; Glasgow also served 61,000 consumers, over a thousand public arc lamps, and nearly three thousand public incandescent ones; also power was supplied to

Govan and Partick from the date that these became part of Glasgow, 1912.

Other systems were: the *Skelmorlie Electric Supply Co* (1911), the *Lanarkshire Hydro-Electric Power Co* (1924—a Clyde Valley and the *Power and Traction Co* system set up to develop the Falls of Clyde scheme) and the tiny *Symington (Lanarkshire) Electrical Supply Co* (1920) which lit the Tinto Hotel at Symington, in the Upper Ward of Lanarkshire. *Rothesay Corporation* (1898) on the Isle of Bute bought power from the *Rothesay Tramway Co*, a subsidiary of the far-flung *British Electric Traction Co.*

South West Scotland demonstrated virtually all the pre-Weir themes of ownership and supply, and the power was generated by water, steam turbines, reciprocating engines, destructor stations and gas or oil engines. Yet what was a glittering store of variety to the collector of the statutorily and technologically arcane was really one of confusion's minor masterpieces. Nowadays the whole area is served uniformly and efficiently by the *South of Scotland Electricity Board.*

Fife

The kingdom of Fife presented a simpler picture, albeit an interesting one. The area was dominated by the *Fife Power Co* (power station at Townhill, near Dunfermline)—itself a wholly owned subsidiary of the *Fife Tramway, Light, and Power Co*, which also owned the Dunfermline tramways and various Orders covering the shire. However, as was so often the case, the company missed some of the richest plums, and was prevented from imposing a truly rational system on its area. *Kirkcaldy Corporation* maintained the tradition of municipal independence with its own power station, distribution network and tramways. The only penetration allowed by the 'lang toun' to the Fife Co's was the granting of certain running powers to the *Wemyss and Dis-*

trict Tramways, an associate of the *Balfour Beatty* group which also controlled the Fife Tramway, Light, and Power Co, and which was, of course, one of the largest electric supply 'combines'. *St Andrew's* electricity supply was in the hands of a smaller group, the *Electric Supply Corporation* (see p 53, and Dumbarton *supra*) and the *Fife Coal Co* ran its own fifteen-mile circuit of power centred on its own power station at Kelty, thus removing another load-building prize from the clutches of Balfour Beatty.

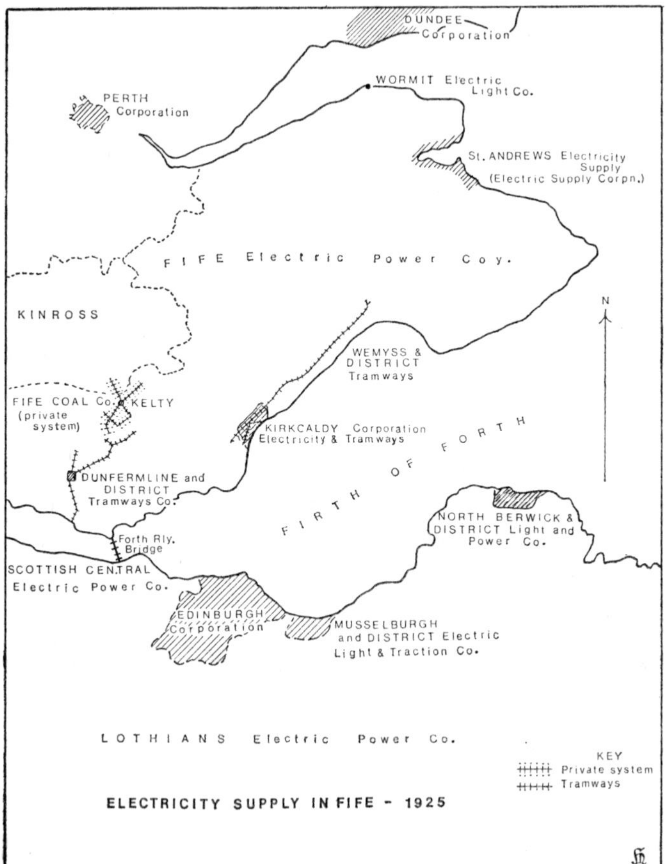

ELECTRICITY SUPPLY IN FIFE - 1925

It is fitting in many ways to end this Appendix, by referring to Fife's other undertaking—the *Wormit Electric Light Co.* This is so for many reasons. Wormit was one of the oldest users of electricity in Great Britain, and was reputedly Scotland's first recipient of public supplies of electricity—offered by the town's developer, Alexander Stewart, at ten shillings (50p) per Quarter, irrespective of consumption. The power was originally generated by a windmill on Wormit Hill (and by a gas engine in later years until the Fife bulk supply arrived in 1930). A symbolic juxtaposition: the oldest power source in the world generating the newest, and looking down upon the *raison d'être* of the little town: the first Tay bridge, a monument to Faustian engineering which collapsed on the very eve of the Electric Revolution: 1879. In all, a timely reminder of our strange and uneasy relationship with technology.

APPENDIX VI: Statistics

(1) Consumers of electricity.

1920	730,000
1929	2,894,000
1938	8,920,000
1970	17,000,000

(2) Electric motors installed in Great Britain

1912	2½ million hp
1924	6½ millon hp
1930	9 million hp
1936	12 million hp

(3) The varying systems, as at 1 January 1930.
Source: Electricity Commissioners' annual returns 1929–30 and 1930–1.

(*a*) *By current:*	*Total*	*Municipal*	*Company*
AC only	236	134	102
AC and DC	274	184	90
DC only	145	49	96

(*b*) *By phase:*

3-phase:	232 stations
2-phase:	16 stations
1-phase:	16 stations

(*c*) Distribution systems by frequency:

Frequency	Number of systems	Frequency	Number of systems	Frequency	Number of systems
25 Hz	35	60 Hz	4	85 Hz	3
30 Hz	1	70 Hz	1	87 Hz	1
$33\frac{1}{3}$ Hz	1	75 Hz	1	90 Hz	1
40 Hz	26	80 Hz	3	93 Hz	1
50 Hz	*	83 Hz	1	100 Hz	9

* standard frequency, generated by 81% of all plant

(4) Ownership of generating plant, 1930:

	Units sold
250 Public statutory undertakers	6,501,191,431
245 Company statutory undertakers	4,415,048,343
28 Railways	826,099,959
20 Tramways	408,197,348
16 Non-statutory undertakers	181,912,763
559 Total	12,332,949,844

APPENDIX VII: (1) Chronology, (2) Nikola Tesla

A number of vital innovations and discoveries made the electric revolution possible. The chief ones were:

(i) *Theoretical basis*

1786	Electrical discharge (Galvani)
1800	Electric current (Volta)
1820	Magnetism associated with electricity (Oersted)
1831	Basic principles of generation and transmission (Faraday)
1873	Magnetism and electricity (J. C. Maxwell)
1897	The electron (J. J. Thomson)
1904	Thermionic valve (Fleming)

(ii) *Generation and transmission*

1832	First dynamo (Pixii)
1860	Lead storage battery (Plante)
1878	Open coil dynamo (Brush)
1880–2	Hydro-electricity (at Rochester, NY, Godalming, Rothbury and Niagara)
1884	Steam turbine (Parsons)
1900–1	British three-phase AC power supply (Merz, Wallsend)

(iii) *Lighting*

1808	First effective arc (Davy)
1848–79	Swan's experiments with incandescent lamps
1877–9	Swan and Edison produce incandescent lamp and 'subdivide the electric light'

(iv) *Motors*

1821	First motor (Faraday)
1863	Separately excited dynamo (Wilde)
1870	First effective dynamo (Gramme)
1879	Electric furnace (Siemens)
1884	AC commutator motor (Siemens)
1887	Electric welding (Elihu Thomson)
1887	Two- and three-phase AC motors (Tesla)
1891	Single-phase AC induction motor (Langdon Davies)
1909	Electrocardiograph
1926+	Improved rectifiers (non-mechanical conversion of AC to DC)

(v) *Traction*

1840	First electric rail vehicle (R. Davidson on Edinburgh & Glasgow Railway)
1883	Electric tramways at Giant's Causeway, and at Richmond Va, USA. The latter pioneered by Frank Julian Sprague, inventor of multiple-unit control and the trolley boom collector

(vi) *Telecommunications*

1833	Electric telegraph (Gauss, Weber)
1894	Radio or wireless telegraphy (Lodge, Marconi)

(vii) *Cables*

1877	Oil-filled cables (Brookes)
1881	Bitumen-covered cables (Callender)

1882	Hopkinson & Edison's three-wire DC supply
1889	Underground conduit (Kennedy)

(2) Nikola Tesla

Although it is not intended to prove, by selective and highly inductive reasoning, that 'all major electrical engineering discoveries were made by 1900', the single example of Nikola Tesla demonstrates how much had been achieved by 1900, and how many possibilities of the new power had been explored in a brief period of brilliant research. The Electric Revolution 'exploded into a vacuum', and covered immense tracts of academic and technological territory in a short time. If 1880–1900 was 'dawn', then to finish the metaphor, it 'came up like thunder' as Kipling had it of daybreak in old Mandalay.

Tesla was a Serbian immigrant into the USA; he had enjoyed an excellent technical education and was an accomplished mathematician—unlike Edison whom he came to scorn. He claimed to have outlined the principles of two- and three-phase generation and distribution in his head in 1882; the actual patent was taken out in 1887. He developed the 'Tesla coil' for getting ultra-high voltages, and shortly after gas-filled and phosphor-coated bulbs that shone without filaments (ancestors of the neon tube)—1890. By 1893 he had produced very high frequency currents, and outlined a complete system of wireless communication. His early flowering culminated with a series of experiments at Colorado Springs (1899–1900) with the largest Tesla coil ever constructed; it was claimed to produce 135 ft long sparks and 12 million volt potentials. He designed wireless guided missiles (1898) and in later years outlined radar (1917). After 1900, however, his wayward genius was rather isolated by a lack of funds and support, not a little owing to his inability to work in a team, for other people and as a member of any community. He started to build a 'World System of Wireless' at a remote spot on Long Island

(1900), but it was never finished. He toyed thereafter with a remarkable series of concepts, including wireless transmission of power; lighting the entire world with a man-made *aurora*; very high tension DC transmission and generation (ironical, in view of his early hostility to DC in general and Edison in particular); a 'death ray' and a 'method of interplanetary communication'. Tesla was very much a lone worker, and highly secretive, so that many of his plans are based on hearsay. He waxed optimistic about his systems, which would be 'man's crowning achievement' and would 'place him beside his Creator'; he could 'cause planets to collide and produce his sun, his stars, his heat and his light'. In all these claims he was a true child of the Electric Revolution, as uncompromising as Ferranti, as far ranging as R. H. Deane. His lack of any commercial or political talent neutralized his technical brilliance and prevented him from developing into a modern Faust. In all these things one sees reflections of the Electric Revolution: an intellectual *tour de force*, slowed by the shifting sands of politics and economics; a disturbing novelty of gigantic potential, hard to contain and employ.

Bibliography

Rollo Appleyard : *Charles Parsons*, 1933

Rollo Appleyard : *History of the IEE*, 1870–1931

H. H. Ballin : *The Organisation of Electricity Supply in Great Britain*, 1946

W. H. Bett and J. C. Gillham : *Great British Tramway Networks*, 1940 and subsequent editions

H. Brearley : *The Development of the Trolleybus*, 1957

I. C. R. Byatt : *The British Electrical Industry 1875–1914* (Oxford University Ph.D Thesis)

D. A. Clarke : *Modern Electric Lamps*, 1952

R. E. B. Crompton : *Reminiscences*, 1928.

P. Dunsheath : *A History of Electrical Engineering*, 1962

The *Electrical Magazine*

The *Electrical Review*

The *Electrical Times*

The *Electrical World*

The *Electrician*

The 'Electrician' *Electrical Trades Directory* and *Annual Tables of Electricity Undertakings*

Electricity Commissioners Reports, 1919–48

Gertrude Z. de Ferranti and Richard Ince : *Life and Letters of S. Z. de Ferranti*, 1934

J. A. Fleming : *Fifty Years of Electricity*, 1921

Garcke's *Manual of Electrical Undertakings and Directory of Officials*

James Greig : *John Hopkinson, Electrical Engineer*, 1970

F. J. G. Haut : *History of the Electric Locomotive*, 1969

J. W. Hammond : *Men and Volts*, 1941

R. A. S. Hennessey : *Power*, 1971 and *The Electric Railway That Never Was, York–Newcastle, 1919* (1970)

Sir Alfred Hopkinson : *Penultimate*, 1930

Hurst and Hazell : *Development of Electric Power Cables*, 1956

E. Jackson-Stevens : *British Electric Tramways*, 1971

R. Jones and O. Marriott : *Anatomy of a Merger*, 1970

J. Joyce : *Trolleybus Trails*, 1963

H. W. A. Linecar : *British Electric Trains*, 1947

Charles Klapper : *The Golden Age of Tramways*, 1961

M. Maclaren : *Rise of the Electrical Industry during the Nineteenth Century* (Princeton, 1943)

W. T. O'Dea : *Social History of Lighting*, 1958

R. H. Parsons : *Early Days of the Power Station Industry*, 1939

T. A. L. Paton and Guthrie Brown : *Power From Water*, 1960

L. E. Perowne : *et al, The History of the Tyne Electrical Engineers, RE, 1884–1933* (1935)

W. L. Randell : *S. Z. de Ferranti*, 1943

Sir Henry Self and Elizabeth Watson : *Electricity Supply in Great Britain*, 1952

Georg Siemens : *History of the House of Siemens*, 1957

Alan Sullivan : *Our Silent Partner*, 1923

A. G. Whyte : *The All-Electric Age*, 1922; *Forty Years of Electrical Progress*, 1930

J. S. Will : *The Law Relating to Electricty Supply*, 1932

Index

Numbers of illustrations are shown in *italic* type.
Electricity supply and distribution companies are listed in the Index and abbreviated thus: E—electric, electrical or electricity; L—light or lighting; P—power; S—supply; T—traction.
Hydro-electric schemes, power stations and railways are listed together.